Computer Tools and Problem Solving in Mathematics

Computer Tools and Problem Solving in Mathematics

James H. Wiebe
California State University, Los Angeles

Terence R. Cannings
Series Editor

FRANKLIN, BEEDLE & ASSOCIATES INCORPORATED
8536 SW St. Helens Drive, Suite D
Wilsonville, Oregon 97070
(503) 682-7668

To Meara, Nathan and Malorie

Microsoft Works, MS-DOS, Microsoft Word and MS Quick Basic are registered trademarks of the Microsoft Corporation.
IBM, PC-DOS and IBM Basic are registered trademarks of International Business Machines Corporation.
Macintosh, AppleSoft and AppleWorks are registered trademarks of Apple Computers, Inc.
Turbo Pascal, Paradox and dBASE IV are registered trademarks of Borland, International.
WordPerfect is a registered trademark of the WordPerfect Corporation.
PageMaker is a registered trademark of Aldus Corporation.

Publisher	Jim Leisy
Interior Design and Production	Lisa Cannon
Editor	David Dexter
Proofreader	Tom Sumner
Cover Photo	© The Stock Market/David Wilhelm, 1990
Manufacturing	Malloy Lithography, Ann Arbor, Michigan

Library of Congress Cataloging-in-Publication Data

Wiebe, James H.
    Computer tools and problem solving in mathematics / James H. Wiebe.
        p.   cm.
    Includes bibliographical references and index.
    ISBN 0-938661-36-1
    1. Mathematics--Study and teaching--Data processing.
2. Mathematics--Computer-assisted instruction.   3. Mathematics-
-Study and teaching--Audio-visual aids.   I. Title.
QA16.W54   1992
510'.71--dc20   92-3375
                                                              CIP

Foreword

The National Council of Teachers of Mathematics (NCTM), during the later half of the 1970s, developed five-year goals for math instruction. These recommendations were published in a booklet entitled *An Agenda for Action: Recommendations for School Mathematics of the 1980s*. The recommendations that the Council stipulated included:

- All students should have access to calculators and increasingly to computers throughout the school mathematics program.
- The use of electronic tools such as calculators and computers should be integrated into the core of mathematics curriculum.
- Curriculum materials that integrate and require the use of the calculator an computer in diverse and imaginative ways should be developed and made available.
- A computer literacy course, familiarizing the student with the role and impact of the computer, should be a part of the general education of every student.

Although calculators are an accepted tool for math instruction and computer literacy courses are fortunately rare in today's classrooms, the Council was very visionary in its recommended use of technology tools.

This book will demonstrate, for teachers at all levels, applications of technology tools for the effective learning and teaching of mathematical skills and concepts as suggested by the NCTM. Application software programs such as word processing, data base management, and spreadsheets have an integral role in teaching mathematics. No longer are math teachers "stuck" with teaching programming classes. No longer is the mathematics department "the" department responsible for teaching computer literacy classes. Teachers, K-12, now have access to a wide range of technology tools that can supplement and complement their teaching skills.

Mathematics can be an exciting adventure! Imagine students in California developing and then comparing family food budgets with students in Japan. Or, imagine students exploring and discovering mathematical concepts through the use of laserdiscs. Today, through the use of technology tools, students can create their own learning microworlds, as well as share data and concepts with the global community of students.

Mathematics can be considered one of the principal problem-solving tools for our society. Mathematics pervades everyone's lives as workers, as consumers, and as citizens. As you develop your own applications of technology, stimulate and encourage your students to explore and learn in different ways. Now that we have the technology tools that can promote alternative learning strategies, we will produce citizens who are equipped academically to be educated participants in today's and tomorrow's world.

Terence R. Cannings, Ed.D., Series Editor
Pepperdine University

Preface

The very name computer bears testimony to the fact that computers were invented to do mathematical tasks. Early computers such as the ENIAC and UNIVAC I were developed in the 1940s and 1950s specifically for the purpose of crunching numbers—doing many mathematical calculations very quickly. One of the most powerful and frequently used computer-based innovations, the spreadsheet, was developed for the purpose of displaying and doing computations on blocks of numerical data. It was not until comparatively recently that computers began to be used extensively for manipulating, storing and printing textual information. As we all know, computers are now used for a vast array of things from controlling the fuel flow in our automobiles to helping composers create music.

As microcomputers have become extensively available in the schools, the variety of uses to which they are put has greatly expanded. They are used by teachers to create all sorts of documents—tests, newsletters for parents, and worksheets, to name a few, and to create posters and graphics for bulletin boards, to keep student records, and to keep track of the materials in and around the classroom.

Students, also, can use computers for many learning activities. They can learn traditional school subjects through a large variety of instructional programs, they can experience simulated real-life situations that would be difficult to bring into the classroom, or they can practice skills learned through traditional instructional activities. Computer tools such as word processors, spreadsheets, and programming languages can be used by students for writing activities, for entering and manipulating data obtained in science and social studies activities, or for exploring number patterns.

In the mathematics classroom, one of the main uses of computers continues be for the processing of numbers: spreadsheets and programming languages can be used to do computations when they are too numerous or complex for calculators.

The graphics components of Logo, BASIC, Pascal, and other programming languages allow the user to explore geometric and spatial concepts. Other packages assist the user to solve or graph equations, and to transform geometric shapes. And, of course, there are numerous good tutorials, simulations, practice programs, and problem-solving environments that can enhance mathematics instruction. Applications such as word processors, desktop publishing programs, graphics packages, and telecommunications can be used by the mathematics teacher for preparing materials for students and managing records. In general, the more powerful use of computer tools like spreadsheets and programming languages by students is for solving mathematical problems, whether for developing a problem-solving plan, organizing and manipulating data, or for carrying out computations.

Computer Tools and Problem Solving in Mathematics focuses on the use of general-purpose computer tools—spreadsheets, data bases, word processors, graphics packages, telecommunications software, and programming languages—in the mathematics classroom. Although *all* the uses of these tools in the mathematics classroom are discussed, the focus is on problem solving. A comprehensive overview of the use of computer tools in problem solving in mathematics and related topics is presented. Chapter 1 introduces general-purpose computer software tools, mathematical problem-solving, and strategies for using computer tools to enhance the problem solving process. Chapters 2 to 6 discuss in detail the use of each type of tool in mathematical problem solving and in other uses in the mathematics classroom. Chapter 7 overviews the use of other technology—calculators, video, and the like—in the mathematics classroom. The text presents numerous suggested pupil activities with computer tools in mathematics. Each chapter also contains questions and laboratory activities for teachers and prospective teachers using the text, and two or more supplementary articles written by other authors that relate to the topics being discussed in the chapter.

This book, in and of itself, is not designed to provide a complete introduction to computer hardware and software, their uses in society, or specific tool software. It is assumed that, prior to or during the course in which this book is used, students will also be introduced to general computer terminology (e.g., *disk drive*, *port*, and other basic terms) and computer applications in society, and receive instruction in and hands-on experience with a specific word processor, a spreadsheet, and a data base manager.

James H. Wiebe
Los Angeles, Ca.

Related Books from FRANKLIN, BEEDLE & ASSOCIATES INCORPORATED

AppleWorks for Students
Patti D. Nogales

AppleWorks for Teachers
Carol McAllister and Patti D. Nogales

DOS 5 Fundamentals
Carolyn Z. Gillay

DOS 5 Principles with Practice
Carolyn Z. Gillay

Introduction to the Personal Computer
Keith Carver and June Carver

Introduction to the Personal Computer using Microsoft Works
Keith Carver and June Carver

Macintosh and You: The Basics
Patricia L. Sullivan

PC/MS DOS Fundamentals
Carolyn Z. Gillay

Quattro Pro 4.0: A Hands-on Introduction to Spreadsheets
Keith Carver and June Carver

Works for Students
Patti D. Nogales

The Technology Age Classroom
Terence R. Cannings and LeRoy Finkel

We Teach with Technology
Greg Kearsley, Mary Furlong and Beverly Hunter

WordPerfect for Windows
Jane Troop and Dale Craig

WordPerfect 5.1: Word Processing to Desktop Publishing
Jane Troop and Dale Craig

Also in the
Using Technology in the Classroom Series
edited by
Terence R. Cannings

Table of Contents

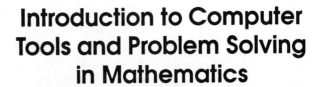

CHAPTER 1

Introduction to Computer Tools and Problem Solving in Mathematics

In the earliest days of microcomputers, there were really only two important types of general-purpose software—BASIC[1] and operating systems[2]. Of course, people soon used these two tools to create programs for doing a variety of very specific tasks, from balancing checkbooks, to keeping track of recipes, to practicing mathematical skills. Then, shortly after microcomputers appeared, more general applications programs started to appear—programs that had mass appeal and which could be used for a large variety of tasks. Spreadsheets, although originally developed for use by accountants, greatly expanded the computer's number-processing capabilities. Word processors turned the computer into a machine that could manipulate words as easily as it could numbers. The data base manager had been around for many years prior to the microcomputer—it was developed for use on mainframe computers in business to keep records on inventory, employees, and the like. However, as microcomputer-based versions of data base programs became available, home and small business users developed many new uses for them. Recently, as computers have invaded our schools and low-cost versions of these originally expensive tools became available, their uses in the classroom have become evident.

Programming languages remain important tools for a variety of purposes. BASIC is used by microcomputer owners who want to do specific tasks that are cumbersome or impossible to do with general applications tools. Many other programming languages, each with its own devoted users and strengths, are available. Logo, Pascal, C, and Cobol are the best-known examples.

---

[1] *BASIC* (Beginner's All Purpose Symbolic Information Code), a computer programming language, has been called "the language of microcomputers," because it comes free or at a minimal cost with most microcomputers

[2] An *operating system* is a set of programs that control the computer. These programs allow communication between the central processing unit and peripherals, allow the user to run programs, save data to disk, copy files, format diskettes, and so on.

Each computer software tool has a variety of uses in the classroom. Teachers can use them for maintaining records or for preparing materials for use by students. Students can use them for preparing high-quality reports and illustrations, for checking their computations done with pencil and paper, or for estimation. But their most important uses are in *solving problems*. These tools give students access to the power of the computer for organizing and manipulating text, data, and shapes in order to help them explore mathematical concepts and solve mathematical problems. An important added benefit is that, when using these packages[3], students are learning to use the types of computer tools used in business and everyday life—the vast majority of home, school, and business uses of microcomputers today involve these general-purpose tools—word processors and desktop publishers, graphics packages, spreadsheets, data base managers, and telecommunications packages. This chapter introduces general applications tools, programming languages, and mathematical problem solving, and overviews the use of computer tools in the classroom.

Introduction to Computer Software Tools

A tool is a device that enhances our ability to accomplish desired tasks. A saw improves our ability to cut wood and a word processor improves our ability to produce and manipulate written text. In this section, we define and present an overview of the uses of general applications tools and programming language packages in the classroom. We also briefly consider computer-assisted instruction packages, even though these packages are usually not considered to be "tool software."

Word Processor and Desktop Publisher

Definition of Word Processor: A word processor is a program which allows the user to input, manipulate, save, retrieve, and print written (typed) text.

Common uses of word processors are to create and print letters, reports, and manuscripts. Fig. 1.1 shows an example of a teacher's letter to parents prepared on a word processor. Note that this letter is set up so that it can be merged with a set of parents' names and addresses—after the letter and the name/address data base have been created, the user requests that they be merged. The word processor then creates one letter for each address with names and addresses in the locations indicated (a last name is substituted for a ^L, a first name for a ^F, and an address for a ^A).

Definition of Desktop Publisher: Desktop publishing programs are designed to create newsletters and similar documents. They are similar to word processors, but allow a much larger variety of type styles (fonts) and formats and allow the user to create graphics in the document or import pre-made graphics from disk and place them in the document.

---

[3] A *software package* is a set of programs on floppy diskette, along with reference manuals, sold as a unit.

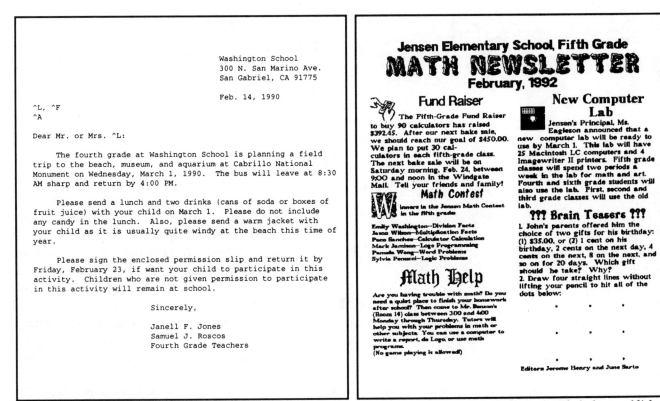

Fig. 1.1: Sample letter prepared on word processor   *Fig. 1.2: Sample document created with desktop publisher*

Desktop publishers (and word processors with desktop publishing capabilities) allow the teacher to create professional-looking, visually attractive handouts. Fig. 1.2 shows a newsletter created with News Master on an IBM-compatible computer.

Most teachers find that, when it comes to classroom management, word processors and desktop publishers are their most valuable computer tools. Because they allow you to type text, modify it in various ways (e.g., move paragraphs, insert text, delete, check and change spelling and grammar), and print it and save it to disk, they greatly facilitate the process of creating and modifying worksheets, assignment sheets, examinations, and other student materials. An important feature of desktop publishers and more advanced word processors, especially for mathematics classroom, are their graphics capabilities. They allow the non-artistic teacher to create pictures, drawings, graphs, or geometric designs in document (or use those created by others). These drawings and designs can be used to model or illustrate mathematical concepts.

Another use of the word processor in the mathematics classroom is to have students use it in problem-solving activities. One of the first steps in the problem-solving process is for students to restate the problem in their own words in an effort to understand what the problem is asking. Using a word processor can motivate students to write more complete restatements of the problem or, perhaps, to restate the problem in a number of different ways. Another useful strategy is to have students write their own problems using a word processor (within certain parameters) to solve themselves or for other children to solve.

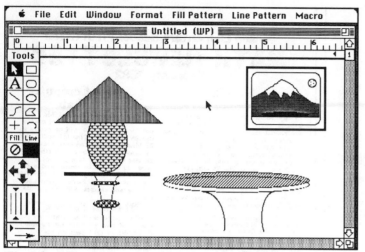

Fig. 1.3 A Graphic created with Microsoft Works

Graphics Package

Definition of Graphics Package: A graphics package contains a variety of tools which allow the user to create designs, pictures, graphs, and other "drawings" on the screen and to print them with a printer.

Typically, graphics packages allow the user to position and size rectangles, ovals, and other regular shapes; draw irregular shapes using a mouse or keystrokes; fill closed shapes in desired colors or shadings; enter text; draw lines of various thickness and colors; and modify previously entered graphics. Beginners can quickly create pictures and designs that are quite interesting and adequate and moderately experienced or talented users can create pictures that, previously, could only have been created by highly talented artists. Fig. 1.3 shows a drawing created with the graphics component of Microsoft Works on a Macintosh computer.

Both teachers and students will find uses for graphics packages in the mathematics classroom. Teachers may use them for creating pictures that model mathematical concepts on worksheets, overhead transparencies, or bulletin board displays. The teacher in Fig. 1.4 is using Microsoft Works on a Macintosh to create a drawing to model decimals. It will be printed out in color on a transparency with an Imagewriter II printer and used on an overhead projector.

Students may use graphics packages in a variety of ways to represent or solve mathematical problems. For example, they may draw pictures or models to help clarify or choose among possible solutions to a problem. Chapter 5 discusses and gives several examples of students' use of graphics packages to solve mathematical problems.

Spreadsheet Package

Definition of Spreadsheet: A spreadsheet program allows users to enter and display numerical data and accompanying labels in rows and columns. It also allows the user to perform calculations on the data in the spreadsheet and display the results.

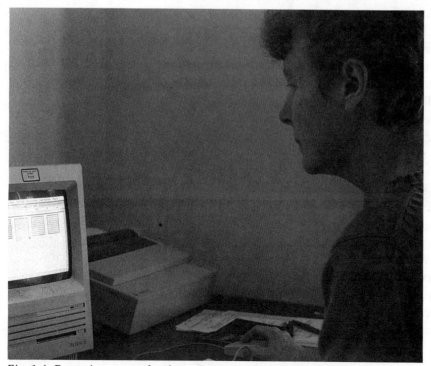

Fig. 1.4: Preparing an overhead transparency with Microsoft Works

Spreadsheet programs, the computerized equivalent to accountants' paper ledgers, were originally developed for the purpose of displaying and manipulating financial information. Like so many other computer tools, however, their use has expanded far beyond the visions of their originators. In the home, for example, they may be used to keep track of income tax records, balance checkbooks, keep track of record or stamp collections, or keep records on telephone or electric usage.

An important feature of computerized spreadsheets is that when you change a value in the spreadsheet (e.g., a student test score), all formulas that use this value (e.g., semester average, total score, or semester grade) are automatically, instantaneously recalculated. This allows the user to do "what-if" types of projections. "What if we decided to purchase the better, more expensive paint—costing $16.95 a gallon instead of $12.95 we had originally budgeted for?" "What would be the total cost of redecorating our school?" "What if we purchased the medium-priced paint at $13.65?" These values can be entered into the spreadsheet and their impact on totals, averages, and the like can be viewed immediately.

Spreadsheets have many uses in the schools and are a valuable tool for solving mathematical problems. Because they allow you to enter numbers and then design your own formula or use one of the built-in formulas to do computations on these numbers, they can be used to investigate number patterns or do quick computations on large blocks of data gathered in problem-solving activities. In addition, most spreadsheets have the capability of presenting results visually in the form of graphs. Fig. 1.5 shows a sample spreadsheet for solving a measurement problem. Chapter 2 in this book will discuss in detail the use of spreadsheets by students to solve problems in the mathematics classroom.

	A	B	C	D	E	F
1	Problem 1: Imagine a set of boxes of different shapes, but					
2	for all of which, the sum of their length, width, and					
3	height is 50 cm. Enter different length/width/height					
4	combinations into the spreadsheet to find out how the volume					
5	and surface area of the boxes change as the dimensions					
6	change. One example is provided below.					
7	Length	Height	Width	L + H + W	Volume	Surface Area
8	25	20	5	50	2500	950
9				50		
10				50		
11				50		
12				50		

Fig. 1.5: Using a spreadsheet to solve measurement problems

The spreadsheet can assist the teacher in a variety of classroom recordkeeping tasks such as grading and monitoring materials and expenses. It is a much more flexible and powerful tool than computer packages designed to do specific tasks such as teacher gradebook programs or inventory control programs. See Fig. 1.6 for an example of a teacher-created gradebook spreadsheet.

File: EMTA1

Page 1
7/12/87

Last Name	First Name	Aug 1	Aug 2	Aug 3	Aug 4	Aug 5	Test 1	Aug 6	Aug 7	Aug 8	Test 2	Aug 9	Aug 10	Test 3	TTL Pts	Avg.
Adams	Jane	9	8	9	9	10	85	6	8	10	88	6	9	94	351	87.750%
Avery	Alvin	6	8	7	9	10	85	8	9	10	90	5	10	87	347	86.750%
Boudreaux	Bernice	8	7	7	8	8	72	8	6	10	80	8	9	95	326	81.500%
Cedera	Juanita		7	6	7	7	78	8	9	9	80	9	9	82	311	77.750%
Chung	Connie	4	6	4	8	6	62	7	8	6	78	10	8	77	284	71.000%
Connand	Barbara	10	10	10	9	10	99	10	10	10	97	8	10	100	393	98.250%
Dovery	Jason	8	9	10	10	7	90	8	9	10	92	10	9	94	366	91.500%
Gradey	Charlotte	10	8	8	9	10	88	7	10	9	89	10	8	90	356	89.000%
Halley	Peter	7	7	5		9	77	6	8	8	75	8	10	79	299	74.750%
Johnson	Jennifer		5		9	10	55	7	5		67	7	8	70	243	60.750%
Keap	George	9	9	9	9	10	92	8	9	9	95	8	9	96	372	93.000%
Landry	Steven	8	10	10	8	9	91	9		10	92	9	10	93	359	89.750%
Luvalle	Marcia	10	10	10	10	10	98	9	10	10	88	10	10	94	379	94.750%
Martinez	Raul	5	8	6	8	9	88	8	9	10	89	7	7	89	343	85.750%
Montinez	Jorge	9	9	9	9	9	97	10	9	9	96	8	9	92	375	93.750%
Oleary	Semantha	8	9	9	9	9	91	9	7	4	82	6	3	67	312	78.000%
Pieau	Piensu		5	7	9	10	88	7	5	10	78	9	9	83	320	80.000%
Quiring	Kalvin	8	9	10	10	10	99	10	7	10	93	9	8	97	380	95.000%
Rousseau	Albert	10	10	10	10	10	99	10	10	10	97	10	10	100	396	99.000%
Vasquez	Jeanne	8	7	10		8	79	9	10	8	82	8	7	85	321	80.250%
Yee	Faith	9	10	9	9	10	93	9	9	10	95	9	10	97	379	94.750%

Fig. 1.6: Spreadsheet with student scores (from Wiebe, 1988, p. 51)

Data Base Manager

Definition of Data Base Manager: A data base manager (DBM) is a program that allows the user to create and manipulate computerized files of information (data bases), and to generate printed reports using that information.

A data base manager is, essentially, a computerized file cabinet. A file cabinet is a data base of information about people, products produced and sold by a company, or events scheduled in a junior high auditorium, for example. Each file folder contains information about an individual person, object, or event in the data base. The problem with a file cabinet is that it is organized in one and only one way—e.g., alphabetical. If you have a file cabinet containing information about all pupils in your school and it is arranged in alphabetical order, it would be very difficult to find all children whose mothers work or all children who live in a certain zip code region.

The computerized data base manager organizes information according to *records* (the equivalent of a file folder). Computerized data bases are much more powerful than file cabinets because they allow you to sort and select based on any of the *fields* (part of the DBM containing a single piece of information) in the data base. For example, even if the first field in your data base is the student's last name, you could quickly select all students who receive free lunches if that was one of the fields in your data base. You can even sort and select on more than one field. If, for example, you had a data base with information about children at the fourth-grade level in your school that included CTBS math scores, *Key Math Diagnostic Arithmetic Test* (Conolly) scores, and number of district math competencies passed, you could quickly find and print all students who scored below the 3.5 grade level on the CTBS Computation subtest and below the 3.3 grade level on the overall *Key Math* score *or* who had passed less than 25 of the district's third grade mathematics competencies. Fig. 1.7 shows one record from a data base created with Microsoft Works on an IBM containing information about a child in a fourth-grade class.

Many data base managers are simple enough for elementary students to use in solving mathematical or integrated math/science or math/social studies problems. Several educational publishing companies also sell pre-prepared data bases of social studies or science information that can be manipulated by students to explore questions in these areas. Chapter 4 in this book discusses the use of data base managers by elementary students for solving mathematical problems.

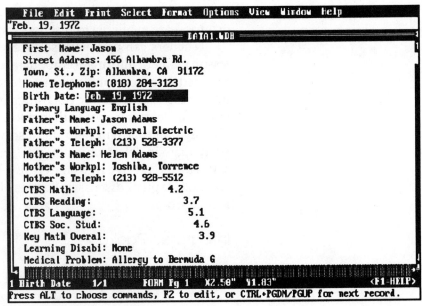

Fig. 1.7: Data base examples

Telecommunications Package

Definition of Telecommunications Package: A telecommunications package contains a set of programs that allow a computer to exchange information with other computers, usually via a modem and telephone lines.

Together with a computer and a modem, a telecommunications package allows the user to communicate with other computers, access on-line services and bulletin boards, and exchange files between dissimilar (or similar) computers. Telecommunications programs convert the electronic signals in a computer to audio tones that can be transmitted across telephone lines. Modem packages will allow you to select and dial telephone numbers from a previously created file of names and numbers, send the characters you type on your keyboard to another computer, receive and display characters from another computer, save received information to disk, send files (e.g., those created on a word processor) to other computers, or print information received (e.g., a file received from a computerized bulletin board).

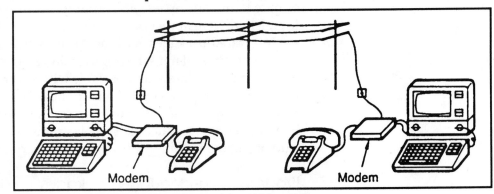

Fig. 1.8: Telecommunications (from Wiebe, 1988, p. 54)

These programs place a vast array of services at the fingertips of the teacher. Commercial services such as Compuserve, The Source, and Prodigy offer on-line shopping, information data-base searches, airline and hotel reservations, and electronic mail. Computer bulletin boards—which offer announcements, information exchange and other services to special interest groups—are available to teachers and students.

Teachers can use these services to obtain or exchange teaching ideas, lesson plans, or information about teacher services (e.g., travel/study groups, reduced prices on computers and teacher materials). Students, also, can tap into these resources for obtaining information about the topics they are studying, learning about other cultures, learning foreign languages, getting information that will help them solve problems, and so on. Many classes participate in "electronic pen-pal" programs, in which children send messages to and receive messages from children in other parts of the country or world. Chapter 7 discusses the uses of telecommunications in the mathematics classroom.

Integrated Package

Definition of Integrated Package: This is a program or group of programs that contain two or more of the above applications and allow easy exchange of data between them.

In integrated packages, the procedures used to accomplish tasks are similar throughout the package, making it easier to learn to use the various programs. Another feature of integrated packages is that data created with one application (e.g., the spreadsheet) can be moved to and used by other applications in the package (e.g., the word processor). Note that many "single application" programs have been expanded to include elements of other applications. For example, many word processors contain certain graphics or desktop publishing capabilities, or simple data bases so that lists of names and addresses can be merged with word-processed documents (e.g., letters to customers).

Integrated packages, such as AppleWorks and Microsoft Works have become very popular in education for the reasons above and because the cost of purchasing an integrated package with several applications is much less than the cost of purchasing each application separately. Teachers can use them for any or all of the management applications discussed above. And, integrated packages such as AppleWorks and Microsoft Works are easy for students in the middle grades and up to learn.

Programming-language Packages

Definition of Programming-language Package: A programming-language package is a program or group of programs that allow the user to enter a set of English-like commands (a program) directing the computer to accomplish a series of tasks. Although the set of commands and syntax understood by programming languages is very restricted, they can be used to accomplish a wide range of tasks.

A programming language is the most general of general-purpose computer tools. Good computer languages allow the user to command the computer to do virtually any task within its capabilities. Although a spreadsheet would allow us to do certain types of mathematical manipulations (namely, perform computations on rows and columns of numbers), there are many mathematical tasks for which they are not of much use. For example, even a relative beginner could write a short program to print out al the prime factors of a composite number, or position a circle of a given size in the upper left corner of the screen, using a programming language. These tasks, however, cannot be done on a spreadsheet.

Typical commands in programming allow the user to direct the computer to print text on the screen or printer:

```
(Logo) PRINT [Hello, my name is Sam.]
(BASIC) PRINT "Hello, my name is Semantha."
(Pascal) Writeln ('Hello, my name is Sylvia.')
```

do a mathematical computation and print the results on the screen,

```
(Logo) PRINT 3*4+5^2.7
(BASIC) PRINT 3*4+5^2.7
(Pascal) Writeln (3*4+5^2.7);
```

assign a value to a variable,

```
(Logo) make "var1 5
(BASIC) var1 = 5  or  LET var1 = 5
(Pascal) var1 := 5;
```

repeat an action a given number of times or until a condition is met:

```
(Logo)4 REPEAT 5 [MAKE "VAR2 :VAR2 + 10 PRINT :VAR2]
(BASIC) FOR VAR2 = 10 TO 50 STEP 10
            PRINT VAR2
        NEXT VAR2
(BASIC)5 DO
            VAR2 = VAR2 + 10
            PRINT VAR2
         LOOP UNTIL VAR2 = 50
(Pascal)6 REPEAT UNTIL VAR2 = 50
            BEGIN
                VAR2 = VAR2 + 10;
                WRITELN (VAR2)
            END;
```

execute a command based on whether or not a particular condition is true or false,

```
(Logo)   IF :ANS = "YES [PRINT [That's good]]
(BASIC)  IF ANS$ = "Yes" THEN PRINT "That's good"
(Pascal) IF ANS = 'Yes' THEN
            Writeln ('That''s Good');
```

[4,5,6] It is assumed that VAR2 has been assigned a value before this routine is executed

or do a large number of other tasks. There are literally hundreds of programming languages available, though only Logo, BASIC, and Pascal are commonly taught in the K-12 classroom. In general, Logo is most commonly taught and used in the primary grades, BASIC in the upper elementary and high school, and Pascal in high school in advanced placement (AP) computer science courses.

Like other general-purpose tools, programming languages have incorporated many features of other applications. Most recently developed programming-language packages for microcomputers have a sophisticated editor that has features found in modern word processors. Increasingly, they come with powerful graphics modules. Turbo Pascal has a graphics toolbox and a module that allows the creation of data bases.

In the classroom, computer programming languages allow the student to do a large range of mathematical explorations. Young children, for example, can use the PRINT statement in Logo or BASIC to check answers to computations done using pencil and paper or mental arithmetic, explore order of operations, or do sequences of computations that would be difficult to do by hand or calculator. They can use looping to do repeated computations, or looping and variables to do computations that make use of intermediate results. Older students can write general procedures or programs using variables and keyboard input to provide solutions to whole sets of problems, explore number patterns, or solve mathematical problems involving complex computations. Students can also use the graphics components of programming languages to explore and solve a large variety of geometric and spatial concepts and problems. For example, children learn much more about regular polygons when writing procedures to draw them in Logo's Turtle Graphics than if they are simply asked to memorize the definitions or identify examples of the different types of regular polygons.

Computer Assisted Instruction (CAI) Package

Definition of Computer Assisted Instruction: A program or set of programs that are designed to deliver instruction to a learner. The categories of CAI are as follows:

Tutorials. These programs act as a tutor or teacher to present concepts or skills to the learner. The user interacts frequently with the program and the program branches to meet the needs of the learner.

Simulations. These present a model or representation of some situation, experience, or phenomenon. They are scaled down to present only the educationally essential components of the event.

Problem-solving software. These present mathematical or spatial problems for the user to solve.

Drill-and-practice software. These programs are designed to help reinforce mathematical skills and concepts taught in another setting. Typically, they provide immediate feedback to the user. They may also provide motivation by embedding the practice in a game setting.

Although this book focuses on general-purpose computer tools such as word processors and spreadsheets rather than on very narrow CAI programs designed to teach or practice specific math concepts to students, it is important to be aware that there are a great many CAI packages designed for the mathematics classroom. They can, in some instances, supplement or provide an alternative to regular classroom instruction.

There are numerous tutorials designed to teach mathematics from the kindergarten level (e.g., Stickybear Shapes, Weekly Reader) through the high school level (Graphing Equations, Sunburst) and beyond. Some focus on number concepts and computational skills (IBM Private Tutor Series, IBM), others on geometry, time, money, and measurement (Clock Works, Money Works, Measure Works; MECC).

Some of the better simulations place children in a problem-solving environment and teach them to apply mathematics to real-life situations. One of the oldest simulations is Lemonade Stand (Apple Corp.), which teaches children to manage a small business and organize money. A powerful, newer simulation is The Voyage of the Mimi. This multi-media package has students solve problems involving science and mathematics.

There is a large variety of good CAI in the area of problem solving. Some programs, such as Discrimination, Attributes, and Rules (Sunburst) or Problem Solving Strategies(MECC) guide students in solving traditional problems. Other packages are designed to develop logical thinking and spatial skills. Some of the better examples are The Factory (Sunburst), Gertrude's Puzzles and Secrets (The Learning Company), Geometric Supposer, and Building Perspective (Sunburst).

Drill-and-practice software for mathematics is the most common type of CAI for microcomputers. Programs that randomly generate math problems, then check to see if the answer is correct are quite easy to develop—indeed, many of the first microcomputer educational programs were developed by math teachers on their kitchen tables (math teachers were often the first to buy microcomputers in the late 1970s). Some of the better, recent programs incorporate practice into a game format. Math Blaster(Davidson), Algebra Blaster(Davidson), Monster Math(MECC), and Math Football (GAMCO).

The use of CAI for developing mathematical concepts, skills, and problem-solving abilities is discussed by Teri Perl in the article at the end of this chapter. She discusses how manipulative materials may be used with children from kindergarten through high school along with computer programs such as The Factory (Sunburst) and Algebra Concepts (Ventura Educational Systems). According to Dr. Perl, "Computers can provide an important link . . . between the concrete manipulatives and the abstract, symbolic, paper-and-pencil representation of the mathematical idea."

The article by Marilyn R. Jussel at the end of this chapter discusses how problem-solving activities involving both computer-assisted instruction and more traditional pencil-and-paper activities were used with fifth- and sixth-grade students.

Introduction to Mathematical Problem Solving

Most of the mathematical situations we encounter on an everyday basis involve problem solving. For example, your favorite department store is having its "Sale of the Century" and you spot a table of designer jeans from

Paris marked down 60%. You find two styles that you would love to buy, but, unfortunately, the sizes are all in centimeters, the dressing rooms are mobbed, the jeans are going fast, and your time is running out on the parking meter. Realizing that accuracy is important (major alterations will be much more costly and probably won't look as good as a tuck here and there), you pull your calculator out of your purse and do the necessary calculations, find the correctly-sized pairs, and head happily to the check-out counter.

Or, you only have $5.00 in your purse and you are having lunch in a restaurant. You decide to order the seafood enchiladas at $3.50. The question is, do you have enough money to order a soda at $.85 and also pay the tax and leave a 15% tip?

In the past, even the recent past, the assumption was made that if we taught students to compute, they would be able to apply their skills when needed. But recent research has shown that having children do pages of computational problems mostly *prepares them to do pages of computational problems*, not necessarily to solve real-life mathematics problems. The most influential organizations involved in mathematics education, such as the National Council of Teachers of Mathematics (1980, 1988), the National Council of Supervisors of Mathematics, and the developers of the California Framework for K-12 Mathematics (1986), have all recommended strongly that we focus our attention away from developing paper-and-pencil computational facility and developing problem-solving capabilities and the ability to apply mathematics in real-life situations. Of course, the deemphasis on pencil-and-paper computations does not mean that we abandon them completely—they will continue to be of some importance, especially in the process of learning mathematics.

Fig. 1.9: Problem-solving situation

So what is problem solving? Does it only involve the ubiquitous "word problem" found in mathematics textbooks? Or is there more to problem solving than that? Before defining problem solving, we need to define what a problem is:

> *Definition of Problem:* A *problem* is a situation demanding resolution for which there is no immediate or apparent solution.

In order for a problem to exist, there must be involvement by the learner, but there can be no obvious solution in sight. Whether or not a problem exists depends not only on the mathematical situation, but the level and involvement of the learner. For example "five plus six" is a problem to most five-year-olds. On the other hand, the following "problem" is not a problem at all to most third graders (it is more accurate to call it a "verbally-stated exercise"): "John had 5 cents. His mother gives him a quarter. How much money does he now have?" A definition of problem solving, then, is as follows:

Definition of Problem -solving: Problem solving is the process of searching for and finding solutions to problems.

Many teachers throw up their hands at the prospect of teaching problem solving. For one thing, it is difficult. True problem solving, by definition, involves a great deal of uncertainty. It is the more or less the opposite of what goes on when children practice computational skills (the process has already been taught, it is simply being practiced—students know what is expected and the sequence of steps involved in the process). Many teachers and students are quite uncomfortable with uncertainty. Another reason why teachers often avoid problem solving is because of the mistaken notion that problem solving is an innate ability—"one either possesses the native ability to solve problems or one doesn't." Another difficulty occurs when teachers use the same evaluation criteria for problem-solving performance as for computation practice. It is most important for teachers to realize that they should not just look at the *product* of the problem-solving activities, they should look at the *process* as well. Children should be given credit for using correct problem-solving strategies (see below), finding the correct resources, selecting appropriate media (e.g., the computer or a calculator), determining whether or not their answer makes sense, partial results, and answers that are reasonably close to the "true" answer. Also, when true problem solving is taking place, one cannot expect the same level of accuracy as with practice activities—70% or 80% is an unreasonably high expectation.

Finally, problem solving does not lend itself to traditional testing. In most instances, problems can and should be solved in a much freer atmosphere than computational drill activities—for example, children might work in pairs or groups; use various reference materials, computational devices, and media; be allowed an indefinite amount of time to find a solution, and so on.

Teachers, researchers, and mathematics educators have found that students' success at solving problems will improve if they solve many problems and are encouraged to use general problem-solving guidelines (heuristics) (Suydam, 1982). George Polya (1973), in his classic book *How to Solve It,* outlines a set of heuristics for solving problems. The first step is to understand the problem. The second is to develop a plan for solving it. The third is to carry out the plan, and the fourth is to look at your answer and see if it is reasonable—if it is not, start over, beginning at step 1 or 2.

Each step in this process may involve several possible substeps. For example, to help them understand the problem, we may ask children to restate (or rewrite) the problem in their own words; to list everything they know about the problem; to make a drawing or a model of the problem or act it out; to check for misunderstandings-pitfalls in the problem or erroneous assumptions they might be making; or to have them identify relevant and irrelevant information in the problem. After the students have demonstrated that they understand the problem, they might be asked to make a rough estimate of the expected answer (Wiebe, 1988).

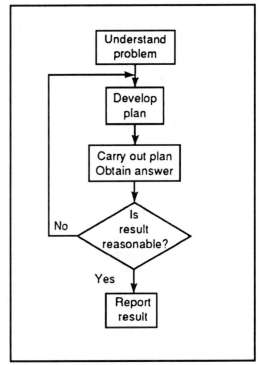

Fig. 1.10: Problem-solving flowchart (from Wiebe, 1988, p. 72).

Step 2 (developing a plan) might involve having the students identify the operations used (and their order) or involved in solving the problem; determine if a table, graph, or diagram will be needed; identify needed tools or resources such as encyclopedias, tables, calculators, computers, or newspapers; and so on.

Step 3 (carry out the plan) would involve actually doing the computation, designing the table, diagram, or graph; looking for patterns; solving a similar but easier (parallel) problem; or writing and solving number sentences, using appropriate computational devices—pencil and paper, calculator, or computer.

Once an answer has been obtained, the student needs to determine whether or not the answer is reasonable. The obtained answer should be compared to the estimated answer (from Step 1). Or the obtained answer can be compared with "expert" answers supplied by the textbook, the teacher, or a group of students. If there is a discrepancy, discuss the difference and, if needed, start the process over again. (Wiebe, 1988)

Some examples are in order:

Problem Example 1

In the auditorium at Washington School, the first 25 rows have 37 seats in each row. The next 5 rows have 39 seats. The balcony has 2 rows with 25 seats and is reserved for teachers. How many students can sit in the auditorium?

Solution:

1. Understand the Problem

   Restate the problem

   The first group of rows in the auditorium has 25 rows and 37 seats. The next group has 5 rows with 39 seats in each. The balcony is reserved for teachers (it has 2 rows with 25 seats each). How many students seats?

   List what you know about the problem:

   25 rows × 37 seats
   5 rows × 39 seats
   (2 rows × 25 seats don't count)

   Check for misunderstandings:

   Don't include seats in the balcony.

   Relevant information:

   25 rows × 37 seats
   5 rows × 39 seats

   Irrelevant information:

   2 rows × 25 seats

2. Develop a Plan:

   Multiply 25 × 37 to get seats in group 1
   Multiply 5 × 39 to get seats in group 2
   Add.

Identify the operations and their order:

Multiply, Multiply, Add.

3. Carry Out the Plan:

The plan might include the use of the PRINT statement in Logo or BASIC
to do the computations:

```
PRINT 25*37 + 5*39
```

4. Does the Answer Make Sense?

The answer I obtained was 1120. This makes sense.

Fig. 1.11: Solution to Problem Example #1

Problem Example 2:

The Super Music Store color codes its compact discs and cassette tapes.
Those marked in red are $13.95, blue are $11.95, yellow are $10.95, green are
$9.95, orange are $8.95, pink are $7.45, black are $6.75, and white are $5.95.
Jenny is president of the sixth grade class and Tom is the secretary. They took
money the class earned in a bake sale to buy a set of CDs for the school dance.
They selected 4 black, 3 white, 1 pink, 5 orange, 7 green, and 1 blue. How
much did they spend?

Solution:

1. Understand the Problem

Restate the Problem

*Jenny and Tom bought CDs for the school dance. They bought
4 marked black, 3 marked white, 1 marked pink, 5 marked
orange, 7 marked green and 1 marked blue. Prices are:
red $13.95, blue $11.95, yellow $10.95, green $9.95, orange $8.95,
pink $7.45, black $6.75 and white $5.95. How much did they
spend?*

List everything you know:

Bought	Price
4 black	$6.75
3 white	$5.95
1 pink	$7.45
5 orange	$8.95
7 green	$9.95
1 blue	$11.95

Relevant information?

Black, white, pink, orange, green, blue CD's were bought.

Irrelevant information?

Red, yellow, not bought. The prices are irrelevant.

2. Develop a Plan

Multiply number of disks of each color by price, then add.

Order of Operations:

Multiply all numbers by their price, add all

3. Carry Out Plan:

```
        A            B            C            D
1   Super Music Store Purchases
2   Solution by Amy Jones
3   * * * * * * * * * * * * * * * * * * * * * * * * * * * * *
4   Color        Price        Number       Total
5   Red          $13.95            0        $ 0.00
6   Blue         $11.95            1        $11.95
7   Yellow       $10.95            0        $ 0.00
8   Green        $ 9.95            7        $69.65
9   Orange       $ 8.95            5        $44.75
10  Pink         $ 7.45            1        $ 7.45
11  Black        $ 6.75            4        $27.00
12  White        $ 5.95            3        $17.85
                               Total       $178.65
```

4. Does the Answer Make Sense?

Yes, $178.65 makes sense.

Fig. 1.12: Solution to Problem Example #2

Problem Example 3:

The seventh grade in Sierra Intermediate School decided to hold a round-robin checkers tournament (everyone plays everyone else in the tournament exactly once). Because of space, time, and equipment limitations, Mr. Jackson, the seventh-grade advisor, told the class that they could have no more than 100 matches. How many kids could participate in the tournament? How many matches did the participants play?

Solution:

1. Understand the Problem
    Restate:

Round robin checkers tournament, with no more than 100 matches. How many kids? How many matches?

    Operations?

Look for patterns.

2. Develop a Plan

Find out how many matches for 1 player, 2 players, 3 players, 4 players, etc. See if there is a pattern.

3. Carry Out the Plan

Solution using graphics package:

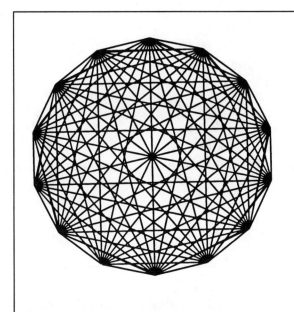

Each dot represents a player. Each line represents a match. Start with two dots and connect them with a line. Each time you add a dot, connect it to all the other dots. Keep track of the number of lines. The figure shows 14 dots (players) and 91 lines (matches). If you add another player, there will be more than 100 matches.

Fig. 1.13: Solution using MacPaint

A possible logo solution:

```
TO TOURNAMENT
MAKE "NUMPLAY 1
MAKE "NUMMATCH 0
    REPEAT 50 [MAKE "NUMMATCH :NUMMATCH + :NUMPLAY
        MAKE "NUMPLAY :NUMPLAY + 1
        PRINT (SE [NUMBER OF PLAYERS: ] :NUMPLAY
        [NUMBER OF MATCHES: ] :NUMMATCH)
        IF :NUMMATCH > 100 [STOP]]
END
```

Note: A Turtle Graphics solution would also be possible, but would be quite cumbersome.

Spreadsheet Solution:

	A	B	C
1	# Players	#Matches	List of Matches
2	1	0	No matches with one person
3	2	1	Player A vs. Player B
4	3	3	A-B; A-C; B-C
5	4	6	A-B; A-C; A-D; B-C; B-D; C-D
6	5	10*	Etc.
7	6	15**	
8	7	21	
9	8	28	
10	9	36	
11	10	45	
12	11	55	
13	12	66	
14	13	78	
15	14	91	The maximum number
16	15	105	of participants is 14

*Pattern: The formula in B6 is +B5 + A5
**Formula from B6 is copied (with relative values to next 25 locations in Col. B)

4.   Does the answer make sense?

Yes, 14 seems reasonable.

Fig. 1.14: Solution to Problem Example #3

Computers Tools and Mathematical Problem Solving

In two of the examples above, spreadsheets provided a convenient tool for listing the data in a table and for doing the computations. Of course, either could have been done using a calculator or pencil and paper. Although it might be a toss-up as to whether or not these two problems merit firing up your computer spreadsheet program and entering the data via the keyboard (as opposed to using a calculator or pencil and paper), such problems could be easily expanded—or similar problems could be developed—to a point where it is clear that the spreadsheet would be the most logical choice. For example, the fourth grade at Washington school won the school contest for the most school spirit. As a prize, the local McDonalds offered one free item from their menu to each of the 83 fourth graders. Seventeen chose Big Macs, 8 ordered large french fries, and so on. What was this prize worth?

Other computer tools can be useful in the problem-solving process. Word processing and desktop publishing programs could be used in all four phases of problem solving, from restating the problem to reporting the answer (see Chapter 5). Graphics programs could be used in step 3, especially if the problem calls for a diagram or a model. Telecommunications programs could be used in problems where on-line data bases would be useful or where information might be gathered from people in other parts of the world (see Chapter 7). Data base programs would be useful when large amounts of alphabetic and numerical data is involved. If a programming language is deemed to be the appropriate tool, the student will need to plan his/her program, write and debug it, and test it to make sure that it is producing the correct or reasonable output.

The process of writing computer programs to solve mathematical (or other) problems parallels general problem-solving strategies (See Chapters 5 and 6). Thus, when students program (with proper guidance), they are learning problem-solving techniques. Computer programming concepts can be used to solve mathematical problems and discover mathematical meanings. For example, a teacher might ask children to write a Logo procedure that "multiplies" two numbers together using repeated addition (i.e., 45 times 75 can be found by first initializing a variable to zero, then using a REPEAT loop that adds 75 to the variable 45 times). Another example is when children are led to discover facts about circles and polygons using Turtle Graphics in Logo.

Another important reason why programming/problem solving activities are beneficial is that programming concepts are used in a variety of other software settings. For example, when using spreadsheets or data base managers, the ability to use conditionals (the IF statement), built-in functions (the square root function), and algebraic formulas (e.g., =(E5+E6+E7)/3) is essential. When children learn to program, regardless of the language, they learn to control input to and output from the computer. They learn how to store information in memory and on disk. They learn to draw pictures on the screen. They learn the value of and applications for mathematics and at the same time they are given a tool to help them solve mathematical and other problems.

Of course, the question remains: When should the computer be used and when should other media be used? If part of your goal is to develop computer literacy and to teach students to use a particular application, you will want to use computers even with problems that could just as easily be solved with other media, as in Examples 2 and 3, above.

Perhaps even more important to your decision is that the availability of the computer allows you to include a variety of realistic (and real) problems with large amounts of data or complex computations in your program—a possibility that did not exist a few years ago. Experts have criticized the problems appearing in math textbooks for being too simple, unrealistic, and narrow in scope. They say that the ability to solve a simple word problem from a textbook has little relationship with the ability to solve real-world math problems. An example of the type of real, challenging, and motivating applied/problem solving activity that elementary school children can solve with the computer is the following:

> Survey local fast-food restaurants to determine how many of selected products are sold at their establishment each month. Based on that information, plan a menu for your own fast-food restaurant.

With the teacher's guidance, the class decides which products and establishments to include in its survey, interviews the restaurant managers to determine how many of each type of product the local fast-food restaurants sold during the month and obtain pricing information about each of the items. The students could then use a spreadsheet to determine the total amount of money spent on each of the products, and (using such programs as Microsoft Works or AppleWorks GS) draw graphs to compare the income obtained from the different products. Finally, they could evaluate the information and plan the menu for their own (imaginary) fast food restaurant! Desktop publishing/graphics programs could be used to print menus. The activity could even be culminated by setting up the "restaurant" and selling food for a day or two to make money for a class party, project, trip, or charity.

Whether you are creating worksheets and examinations for your students with a word processor, keeping track of the materials in your classroom with a data base, or creating posters with a desktop publisher, or whether your students are using a spreadsheet to organize data in order to answer a mathematical problem, or using a programming language to draw mathematical shapes on the screen, you will find that computer tools enhance your teaching of mathematics.

Bibliography

An Agenda for Action: 1980. Recommendations for School Mathematics of the 1980s. 1980. Reston, Va.: National Council of Teachers of Mathematics.

Carpenter, T. P. et al. 1988. Results of the fourth NAEP Assessment of Mathematics: Trends and Conclusions. *Arithmetic Teacher*, 36(4, December) 38-41.

Conolly, A.J., Nuchtman, W., and Prichett, E.M. 1971. *Key Math Diagnostic Arithmetic Test.* Circle Pines, Minn.: American Guidance Service.

Curriculum and Evaluation Standards for School Mathematics. 1988. Reston, Va.: National Council of Teachers of Mathematics.

Johnston, J. 1987. *Electronic Learning: From Audiotape to Videodisc.* Hillsdale, N.J.:Lawrence Erlbaum.

Krulik, S. and Reys, R. E. (Eds.). 1980. *Problem Solving in School Mathematics: 1980 Yearbook of the National Council of Teachers of Mathematics.* Reston, Va.: National Council of Teachers of Mathematics.

Mathematics Framework for California Public Schools, Kindergarten through Grade Twelve. 1986, California State Department of Education.

National Association of Supervisors of Mathematics. 1989. Essential mathematics for the twenty-first century: The position of the National Council of Supervisors of Mathematics, *Arithmetic Teacher*, 36(9, May) 27-29.

Papert, S. 1980. *Mindstorms: Children, Computers and Powerful Ideas.* New York: Basic Books.

Polya, G. 1973. *How to Solve It, 3rd Edition.* Princeton, N.J.: Princeton University Press.

Suydam, M. 1982. Update on research on problem solving: Implications for classroom teaching. *Arithmetic Teacher* 29(6, February): 56-60.

Wiebe, J. H. 1987. Computer Software. In Lindsey, J., Ed. *Computers and Exceptional Individuals.* Boston: Merrill.

Wiebe, J. H. 1988. *Teaching Elementary Mathematics in a Technological Age.* Scottsdale, Ariz: Gorsuch Scarisbrick Publishers.

Wiebe, J. H. 1990. Teaching mathematics with technology: Data bases in the mathematics classroom. *Arithmetic Teacher*, 37(5, January): 38-40.

Discussion Questions

1. What capabilities of spreadsheets are also shared by data base managers? How are these two tools different?

2. Which application would be most useful for you as a teacher in managing your classroom, a word processor or a desktop publisher? Why?

3. What does a child learn about computers when using a word processor or spreadsheet that s/he does not learn about computers when using CAI?

4. What were your experiences with and attitudes toward problem solving when you were in the elementary grades?

5. List the five most recent times you have used mathematics in everyday life (excluding experiences related to classes you are taking or teaching). Classify each as *problem solving* or *routine calculation*. For each problem-solving situation, justify why you labeled it problem solving and not routine computation.

6. Write two problems for students at the grade level you teach (or plan to teach) that would best be solved using a computer tool.

7. In what types of problems would spreadsheets be more appropriate tools than programming languages?

Activities

Activity 1.1

Objective
To find out how teachers and non teachers use computers, especially general applications packages.

Rationale
Discussions with regular computer users will help the student focus on the power of computer application tools in doing everyday tasks, both at home and for the classroom.

Materials
Questionnaire.

Advance Preparation
Read Chapter 1. If the tools discussed in this chapter are unfamiliar to you, have someone demonstrate them to you and, if possible, get some introductory hands-on experience with them.

Procedures
You will use the questionnaire below to interview a teacher who regularly uses computers to manage his/her teaching and a nonteacher who regularly uses a home computer to manage his/her finances, create documents, and the like.

Questionnaire for Interview
1. How do you use computers?
2. Which of the following do you own or have access to?
   a. Word Processor
   b. Desktop Publisher
   c. Spreadsheet
   d. Data Base Manager
   e. Graphics Package
   f. Telecommunications Package
   g. Integrated Package
   h. Programming Language
3. Which of them do you use and how do you use them?
4. How did you learn to use these packages?
5. Of those you don't have, which would you like to have?
How would you use them?
6. How do you envision yourself using computers in the next few years?
7. Which of the above computer applications do you think that you would not use, even if they were made available to you? Why?

Follow-up Discussion Questions
1. How did the answers of the teacher and nonteacher compare?
2. Which packages are most frequently used by the interviewees?
3. Which packages are they most interested in obtaining?
4. Which do they have little interest in using?
5. What do you conclude about the value of the different applications discussed in this chapter for your personal and professional use?

Activity 1.2

Objective
Use a word processor to write a letter to an administrator.

Rationale
It is important for professionals to be able to prepare a well-presented, professional-appearing case for needs and desires. The word processor can be of great help in this aim.

Materials
A computer, word processor, and printer. A catalog from a teacher supply retail or mail-order store.

Advance Preparation
If you have not used a word processor before, obtain a tutorial that introduces the basic word-processor functions to you. Learn how to enter text, insert, delete, use the tab function, underline, move text, and make a printout.

Procedure
Assume that you are teaching in a regular classroom in a K-12 school. Also assume that the school district has money available to purchase materials for math programs. However, to obtain this money, you must write a request specifying the materials you would like to get and how they will be used in your program. Using a word processor, write a letter to your principal describing and listing at least three different materials you would like to purchase from a teacher-supply store or mail-order company. Include complete product information so that the materials could actually be ordered. Also, write a rationale for each of the materials you choose. Make a printout of the letter.

Activity 1.3

Objective
Identify problems from math textbooks.

Rationale
The purpose is to improve awareness of problem solving in elementary mathematics.

Materials
An mathematics methods textbook series.

Advance Preparation
Read Chapter 1.

Procedure
Mathematical problems (verbally stated) that involve simple computations, such as adding two or three numbers, are often called *routine problems;* those that involve more complex computations (multiple operations or the kind not typically practiced in mathematics classes) or logical or graphical solutions such as Problem Example #3 above are called *nonroutine problems.* Find a mathematics textbook written for upper elementary, junior high or high school (depending on the level you plan to teach) and identify one nonroutine and two routine problems from the book.

Activity 1.4

Objective
Use problem-solving strategies.

Rationale
Teachers can get a better understanding of the problem-solving process by actually solving problems themselves.

Materials
Pencil, paper, computer, spreadsheet package, calculator.

Advance Preparation
Read Chapter 2. Possible preparation: become familiar with a computerized spreadsheet program (for those not ready to use a spreadsheet, nonspreadsheet solutions are possible).

Procedure
Use appropriate problem-solving strategies to solve the following problem:

Jesse inherited a coin collection from his great-grandfather who had tried to get a complete set of U.S. coins—a penny, a nickel, a dime, a quarter, a half-dollar and a silver dollar—for each year from 1920 to 1943. Many coins were missing, however, and Jesse vowed that he would complete the first five years of the collection during the coming year, the next five years during the following year, and so on. Jesse has the following coins for the 1920 to 1924 period—1921 quarter, 1923 dollar, 1920 penny, 1920 dime, 1924 dollar, 1922 dime, 1921 nickel, 1922 penny, 1923 penny, 1924 quarter, 1922 half-dollar, 1923 nickel, 1920 quarter, 1920 dollar, 1923 half-dollar, 1921 dollar, 1924 dime, 1924 penny, 1921 dime, 1922 quarter, 1924 half-dollar, and 1923 dime. Which coins does he need to get to complete his collection for 1920 to 1924? What is the face value of all the coins he already has for 1920 to 1924? How much will it cost him to complete the collection if silver dollars from that period cost $15.00, half-dollars cost $8.50, quarters cost $4.75, dimes cost $1.75, nickels cost $.73, and pennies cost $.35?

Activity 1.5

Objective
Evaluate a CAI package for the mathematics classroom.

Rationale
There are a number of generally agreed-upon criteria used in evaluating educational software. Software evaluation forms or checklists that include these criteria will help make the process of selecting educational software more formal, consistent, and likely to satisfy your needs.

Materials
Software Evaluation Form (Fig. 1.15), pencil, paper, computer, CAI software for the mathematics classroom.

Advance Preparation
Read the section on CAI in this chapter.

Procedure
Obtain a CAI software package designed for the mathematics classroom, preferably one that is designed to teach a topic you consider to be important for the math classroom and at the grade level you teach or plan to teach. Spend a half-hour or more with the program, going through as much of it as possible. Evaluate it using the Software Evaluation Form in Fig. 1.15.

Software Evaluation Form

Program Name_____
Publisher_____
Product Number_____Price_____
Subject Area_____
Topic Covered_____
Grade or Ability Level_____
Peripheral Devices Needed_____
    Printer_____   Joy Stick_____   Color Monitor_____
    Mouse_____   External Speaker_____

Type of Program
_____Tutorial (Didactic?_____ Discovery?_____)
_____Drill and Practice (Game Format?_____)
_____Simulation
_____Problem Solving/Logical Thinking

Program Evaluation
    Indicate the level to which the program meets the following criteria:
        A — Program does an excellent job
        B — Above average
        C — Average
        D — Below average
        E — Program does a poor job

A B C D E   Screen presentation is clear, uncluttered, appropriate, and motivational
A B C D E   Graphics are clear and appropriate to instruction
A B C D E   The student controls the pace of instruction
A B C D E   Reading level is appropriate
A B C D E   Program is easy to use, does not require extensive explanation, does not require frequent reference to external documentation
A B C D E   Program provides on-screen help, if needed
A B C D E   Valid instructional techniques used to accomplish the desired instructional objectives
A B C D E   Content is accurate
A B C D E   Program provides feedback to the student
A B C D E   Program meets its stated objectives

Overall Rating
    Outstanding   Excellent   Good   Fair   Poor

Comments

Fig. 1.15: Software Evaluation Form (modified from Wiebe, 1988, pp. 61-62).

Manipulatives and the Computer: A Powerful Partnership for Learners of All Ages

by Teri Perl

Children are sitting on the floor in small groups. In the middle of each group is a set of brightly colored geometric shapes called attribute blocks. Look carefully and you see four shapes, each in three different colors and two different sizes.

The game the children are playing involves building a "train" of blocks that differ from each other by only one characteristic at a time. A child stares at a large blue triangle, the last piece in the train. Several pieces can be placed next: a large blue square, small blue triangle, large yellow triangle, and so on—but not a yellow circle or red square, because they differ in more than one way.

The children are engrossed in their games. The teacher walks around and notices that some students are building trains that don't follow the rules. They are happily building their trains, but some of the adjacent shapes are different in more than one way.

Tomorrow the teacher will introduce these games on the computer using *Gertrude's Secrets* (from The Learning Company). It won't be necessary to monitor the students so closely; the computer will do that. This is one of the advantages of using manipulatives and computers together. We will be describing others in the pages that follow.

What Are Manipulatives and Why Do We Need Them?

Manipulatives are physical objects that can be picked up, turned, rearranged, and collected. They leave no messy trail as a pen does when an answer is changed or alternate solutions are explored.

Manipulatives are useful for modeling ideas. For some time now they have been used in math labs to help children understand abstract concepts. For example, base-10 blocks are used to teach our number system, and how it is possible to name any number—no matter how large—using only ten digits.

Support for the use of manipulatives comes from Piagetian theory, in which cognitive development is described as moving from concrete to abstract, through a series of developmental stages that are roughly age-related. A concrete-operational child cannot handle abstract concepts before arriving at the appropriate stage. However, with manipulatives it is possible for such a student to take the first steps towards exploring the concepts; manipulatives are concrete introductions to abstract ideas.

Much research supports the value of manipulatives in learning mathematics. And, contrary to the belief that manipulatives are useful only for little kids, these results hold up for older students as well. In a summary article *(Research on Instructional Materials for Mathematics,* ERIC/SMEAC Special Digest No. 3,1985), Marilyn Suydam reports, "The use of manipulative materials in mathematics instruction results in increased achievement across a variety of topics at every grade level from kindergarten through grade 8, at every achievement level, and at every ability level."

Manipulatives and Software— Another Link in the Chain

As the evidence supporting the educational value of manipulatives continues to build, more and more educators are incorporating manipulatives into the math curriculum. At the same time, computers are finding their way into classrooms in increasing numbers. This raises questions about how computers and manipulatives relate. As we introduce computer-based learning, are we in danger of giving up the power of manipulatives? Is there anything that computer software can add to the process?

The National Council of Teachers of Mathematics, the main professional organization of math teachers, identifies an important role for computers "as tools to assist students with the exploration and discovery of concepts, with the transition from concrete experiences to abstract mathematical ideas." Computers can provide an important link in the chain, a connection between the concrete manipulatives and the abstract, symbolic, paper-and-pencil representation of the mathematical idea.

Researchers have substantiated the value of using manipulatives and computers together. For example, in a paper delivered at the International Congress of Mathematics Education in Budapest in 1988, Judith Olson reported on research with third-graders who used classification materials such as attribute blocks followed by software featuring similar activities. Her conclusion: Students who used both manipulatives and computer software demonstrated much greater sophistication in classification and logical thinking than did a control group working only with manipulatives.

Another more informal study, conducted in 1988 for Sunburst Communications by M. Bieck and M. Wilson, also concluded that software and manipulatives together can be more effective than either used alone. Third- and fourth-graders at Park Dale Lane Elementary School in Encinitas, CA, were divided into two groups, one using *Puzzle Tanks,* the other using water and containers to solve the same problems. The report states, "Students pouring real water gained a better understanding of the step-by-step processes involved through enforced record keeping, while students using the software concentrated on finding ways to solve the problems and solved the problems faster."

Support for the view that manipulatives and software can be a powerful combination also comes from constructivism, an approach to the study of how learning takes place. As summarized by Carol Maher, a leading thinker in the field, "The constructivist perspective on learning suggests that knowledge is not received passively; it has to be built up. Learning is an act of construction. It is important for learners to have the opportunity to build up a number of representations or models (both concrete and imagistic) and then to make the connection between the different modes of representation." By offering students the opportunity to work with manipulatives, pencil and paper, *and* software, we enable them to build a greater number of models and to begin making such connections.

If we are convinced that both software and manipulatives have a role in improving math learning and that the two together are more powerful than either alone, it's time to look at some of the products on the market that make such a combined approach possible. In attempting to pair manipulatives with complementary software, there are at least four different models to think about. Let's call these *mirroring, modeling, manipulating,* and *managing.*

Software as Mirror

Mirroring is one of the most straightforward ways in which software can complement manipulatives. In this model, the software displays objects that look like corresponding manipulatives and are used in exactly the same way.

The scenario presented earlier, in which students used attribute blocks and then *Gertrude's Secrets,* is an example of mirroring. The screen objects in *Gertrude's Secrets* are easily matched with the attribute blocks children hold in their hands. Both on screen and off, children engage in virtually identical activities, "picking up" the shapes and moving them

around. Similarly, *Moptown Parade* and *Moptown Hotel* mirror activities students can engage in away from the computer with "people pieces" (blocks similar to attribute blocks, but in the shape of people).

Other examples of mirroring software include *ElasticLines* (designed by Education Development and published by Sunburst), which allows students to perform geoboard activities at the computer; *Puzzle Tanks* (an older Sunburst program created by Dr. Thomas C. O'Brien), which challenges students to identify a specific quantity of liquid by "filling and emptying" two on-screen containers of different sizes as many times as desired; and *Algebra Concepts* (from Ventura Educational Systems), which includes activities based on algebra tiles to help junior high and high school students understand operations with polynomials.

In addition, there are programs such as Warren Crown's *Math Concepts* and EDC's *Exploring Measurement, Time, and Money* (both published by IBM); and *Hands-On Math* (from Ventura) that take a more comprehensive approach, mirroring a large number of manipulatives commonly found in the math classroom.

Advantages of Mirroring

Since mirroring software allows students to perform virtually the same tasks on and off the computer, one might wonder what the software adds to the process. In the mirroring model, software enhances the manipulatives in several important ways.

• *Software bridges levels of abstraction.* No matter how effectively software mirrors actual manipulatives, the screen representation is more abstract than three-dimensional models. As noted above, many educators believe that this extra level of abstraction provides an important bridge between the concrete manipulatives and the more symbolic mathematical constructs. For example, Barbara Bayha, former lead teacher in Apple's Classroom of Tomorrow's Cupertino site, finds it useful when teaching about grouping and patterns to allow primary students to progress from the people pieces manipulative to *Moptown* software to letter and number patterns on paper.

• *Software automatically monitors the activities.* Most good problem-solving activities, like the attribute games described earlier, allow a range of "correct" answers. This latitude makes the activities difficult to monitor, particularly by one teacher with a class of standard size. The computer, on the other hand, makes certain that the children's moves follow the rules. It can provide feedback in other ways, too. For example, in *Measurement, Time, and Money,* the user can drag coin combinations on the screen and hear, through digitized speech, the total sum of the set created. With the computer's help, children are less likely to build inaccurate conceptions. And at the same time, the software provides the "pat on the back" that encourages children to go on.

• *Software provides an "endless" and flexible supply of materials.* In certain situations, software plays an important role by giving students access to many more manipulative pieces than would be feasible without the computer. For example, *Math Concepts* offers an unlimited supply of unit cubes, ten-cube long rods, and 100-unit flats, making it possible for a student to build a representation of a very large number without worrying whether there will be enough cubes to go around.

Other times, the computer helps by providing a vast number of *choices.* For example, in *Puzzle Tanks* the computer generates a broad assortment of container sizes for students to use when "pouring" liquid back and forth; and with *Elastic Lines* students can create a geoboard with dimensions ranging from five to 15 pins across, in a square, triangular, or circular pattern. While it is possible for a teacher to purchase or scrounge actual containers and geoboards to represent such a vast range of choices, it is certainly not easy.

• *Software can extend the capabilities of the manipulative.* In a number of cases, the software goes beyond exact mirroring, adding an extra level of interaction. In *Elastic Lines,* for example, students can copy shapes they've created on the geoboard to a special drawing area and use computerized tools to flip and manipulate the shapes in ways not possible with a geoboard alone.

In *Gertrude's Secrets* we see another example of software extending the range of manipulatives. Here users can transform the standard geometric shapes into new shapes, or design their own. Playing the same games with different shapes significantly refreshes and enriches the activities.

• *Software is easier to manage and to clean up.* There are times when a teacher might prefer software to the equivalent manipulative because of the setup and cleanup time saved. For example, using the *Puzzle Tanks* program is easier and less messy than locating a sink, filling containers with water, carrying them back to the classroom, and pouring the water from one container to another. Even with manipulatives that create less of a mess, the software can save valuable time. Turn off the computer and the manipulatives vanish; turn it on again and they reappear in seconds. This is a clear benefit with materials that have many pieces, such as attribute blocks or base-10 blocks.

• *Software (with a projection device) makes it easier to demonstrate concepts to a large group.* Many teachers find it useful to teach large-group lessons with the help of manipulatives. For this reason, a number of manipulative providers sell two-dimensional representations that can be placed on an overhead projector and used for demonstrations. With the computer and an LCD panel such as the PC Viewer, this process becomes easier and more powerful. For example, with *Algebra Concepts,* a teacher can quickly produce an arrangement of algebra tiles during a whole-class presentation without having to count out each tile and arrange it in place on the overhead projector.

The Drawbacks of Mirroring

Of course, there are also disadvantages to using the computer. Some mirroring programs on the market can be more bothersome than enlightening. It is not clear, for example, that online tutorials on the use of manipulatives offer much advantage over printed instructions. While the computer can enhance the use of manipulatives, something is generally sacrificed as well in the transfer from table-top to screen. Even in the best examples of mirroring, computer users must invest time in learning how to manipulate objects that require no explanation away from the computer. (For example, remembering which key to press to lift or place a rubber band on the geoboard in *ElasticLines* is far more confusing than the same task off-line.)

Further, when graphics (or hardware) are not state-of-the-art, the speed at which manipulatives are drawn, erased, and redrawn can be agonizingly slow. Fortunately, however, in many of the software products on the market, the limitations are more than offset by the advantages the computer offers.

The Manipulative as Model

A second way in which software and manipulatives work together can be referred to as "modeling." In this approach, the software plays the central role and the manipulative serves as an aid to visualizing the concepts developed at the computer.

For example, users of *The Factory* (designed by Marge Kosel and Mike Fish of Sunburst) are challenged to create a "product" by designating rotations, punches, and stripes to duplicate a given object. The sequence in which the operations are performed is crucial and can be difficult to visualize without off-line help. A simple paper square can be most helpful. "When I introduce *The Factory* to my class," says Diane Resek of San Francisco State University, "I first hand out squares to each student. Then they can manipulate their squares to act out a sequence of moves to help them solve the problem on the screen."

Manipulatives are helpful for users of another Sunburst program as well. *Building Perspective* (designed by Thomas C. Bretl) asks students to identify the relative heights of a series of buildings, given a top-down view and access to front, back, and side views. I found the task close to impossible until I brought out my set of Cuisenaire rods to keep track of the information offered by the different views. While the goal is clearly to develop the spatial visualization skills needed to complete the puzzles *without* added help, the manipulatives serve as an important first step to understanding and visualizing the problem.

The Computer as Manipulative

A new class of software has emerged in which the software itself serves as a manipulative. In this approach (made possible by new, more advanced graphical interfaces), the online representation rather than an off-line model is what users "touch," stretch, move, etc.

As substantiated by a recent study (reported by Cynthia Char at the 1989 AERA conference in San Francisco), "Software can provide a new manipulative environment for young children's mathematical exploration, offering comparable and sometimes increased user control and flexibility over real hands-on materials." Such observations hold for older users as well. In fact, the computer's role as a manipulative is particularly important when teaching higher-level topics difficult to simulate away from the computer. For example, with *Calculus* (developed by Sensei for Brøderbund), students drag a mouse along a curve and watch the slope-line appear at each point the mouse touches. Simultaneously, an algebraic expression in a window on screen changes to reflect the new curve. The old system of laboriously calculating and plotting points by hand is beautifully taken over by the computer. In *Geometry,* another program in the series, there are other opportunities for students to touch, stretch, and manipulate shapes and figures onscreen, making inferences and discoveries as they do.

Writing about his current research on the value of manipulatives to help students understand concepts in algebra (*The Situated Activities of Learning and Knowing Mathematics,* Stanford University and the Institute for Research on Learning, 1988), Jim Greeno states, "If situations can be found in which students reason easily and effectively about functions and variables, these situations could be used to anchor students' understanding of algebraic notation. We might say, then, that this research could result in identifying materials for algebra that play the role that place-value blocks, Cuisenaire rods, and fraction pie diagrams play for more elementary mathematics."

Such products are on the way. In fact, Warren Robinett's *Snap & Click Math,* in the final stages of development when we went to press, should soon be available from Tom Snyder Productions. This program is a construction set for building animated mathematical machines that move and compute. Users are able to view operations or equations in several different forms (a tree diagram, a standard equation, etc.), take the model apart, explore its components, then put it back together and watch how values flow through it.

Another product that provides online manipulative experiences not possible without a computer is *Letterforms and Illusions* (by Scott Kim, published by W.H. Freeman). These unusual puzzles build spatial visualization skills and offer experience with flips, rotations, and other elements of transformational geometry. Players create and decode messages using a number of cleverly designed fonts. One, for example, allows each letter to be broken into straight and curved lines; another is made up of letters that stretch into sinewy patterns; and so on.

The Software as Manager

Managing refers to a model in which the software and the manipulative work together as a single product. with the software controlling the manipulative. *LEGO TC Logo* is a wonderful example of this sort of relationship. Here the manipulative, in the form of a machine or robot-like device, is managed, maneuvered, and manipulated under computer control.

With *LEGO TC Logo,* users build off-line machines; connect them with wires and an interface box to a computer that speaks a dialect of *LogoWriter;* and enter simple commands to turn motors on and off, reverse their direction, or respond to feedback from specially mounted touch sensors and light sensors. This relationship between manipulative and software is particularly relevant since it gives users an experience of a real-world situation where machine performance is controlled by computer software.

Looking Ahead

Manipulatives and software are individually important for math teaching, and are particularly valuable when used together. Researchers, publishers, and educators all have an important role to play in spreading the word about such a partnership. Researchers can continue to study the effects of the combined use of manipulatives and software on learning, and report their findings to the educational community. Publishers can continue to develop programs in which the software itself is the manipulative and can begin offering packages that bundle "matching" software and manipulatives, making it easier for schools to buy them together.

And finally, teachers can become more cognizant of the connections between manipulatives and computer software. Understanding how these two classes of educational tools work together will allow us to maximize the benefits of both, using them to improve the teaching of mathematics in elementary and secondary schools throughout the country.

Ordering Information for Software Mentioned in This Article

Brøderbund Software
PO Box 12947, San Rafael, CA 94913-2947;
(800) 521-6263
• *Calculus*: Macintosh (512K) School
  edition, $109.95
• *Geometry*: Apple IIgs (512K) school edition $89.95
  Macintosh (512K) school edition, $109.95.

IBM Educational Software
PC Software, 1 Culver Rd., Dayton, NJ 08810;
(800) IBM-2468
• *Exploring Measurement, Time and Money*.
  IBM PS/2 only (uses mouse and speech adapter).
  School price, $136.
• *The IBM Math Concepts Series*. IBM PC, PS/2.
  Level P $81; levels 1&2, $95 each; levels 3&4,
  $106 each.

LEGO Dacta
555 Taylor Rd. Enfield, CT 06082
 (800) 527-8339 or (203) 749-2291
• *LEGO TC Logo*: Apple IIe, IIgs, or Apple II family,
  IBM PC, Tandy 1000, $485 each.

The Learning Company
School Division, 6493 Kaiser Dr.,
Fremont, CA 94555-3612;
 (800) 852-2255
• *Gertrude's Puzzles* and *Gertrude's Secrets*:
  Apple II family; IBM PC, Tandy 1000.
  School editions, $59.95.
• Moptown Hotel and Moptown Parade:
  Apple II family (48K). School editions, $59.95.

Sunburst Communications
101 Castleton St., Pleasantville, NY 10570;
(800) 628-8897
• *Building Perspective*. Apple II family (64K);
  IBM PC, PCjr, PS/2 (Model 125), Tandy 1000;
  Commodore 64. $65 each.
• *The Factory*. Apple II family (64K) IBM PC,
  PCjr, Tandy 1000; Commodore 64; TRS-80;
  $65 each.

Tom Snyder Productions, Inc.
90 Sherman St., Cambridge, MA 02140;
(800) 342-0236 or (617) 876-4433
• *Snap & Click Math*: Macintosh (512K).
(Not available at press time. Contact the
publisher for information.)

Ventura Educational Systems
3440 Broken Hill St., Newbury Park, CA 91320;
 (800) 336-1022 or (805) 499-1407 in CA
• *Algebra Concepts*: Apple II family (64K);
  IBM PC, Tandy 1000 (512K); Macintosh (512K).
  $49.95 each.
• *Hands on Math*: volumes I-III: Apple II family
  (64K). $49.95 each.

W.H. Freeman (to the attention of Susan Perpeluk)
41 Madison Ave., New York, NY 10010,
(801) 973-4660
• *Letterforms and Illusion*: Macintosh
  (requires MacPaint). $39.95.

Teaching Problem-solving Strategies With and Without the Computer

by Marilyn R. Jussel

The question many educators are asking today is how to effectively teach problem solving. Can computers provide an answer? The development of problem solving skills is an important objective in education, but it is also a difficult objective for students to achieve and for teachers to teach. Recent research indicates that to improve students' problem-solving skills, specific attention must be given to teaching problem solving strategies. According to LeBlanc, Proudfit, and Putt (1980), "Success in solving the problem does not depend on the application of specific mathematical concepts, formulas, or algorithms; rather, the solution requires the use of one or more strategies" (p. 105). Studies indicate that "students need to have explicit instructions in how to use strategies, to see examples of strategies being applied to solve problems, and to have opportunities to practice problem solving" (Truckson, 1982/3, p. 7). Hence, the student must be aware of the process in order to solve the problem.

Our project, supported by a "Teachers Can't Wait" Apple Education Foundation Grant and Sunburst Communications, Inc., combined problem-solving strategies and computer-assisted instruction. We attempted to match the problem-solving skills needed for the future with the technology that will be an integral part of that future. Instructors from Kearney State College and Holdrege Middle School reviewed software and developed a curriculum that matched specific problem-solving strategies with computer software. The results of this pilot study and a subsequent more structured study (Jussel, 1988) showed that the computer can be successfully used to help students develop their problem-solving skills.

The curriculum, developed for fifth and sixth grade students, uses materials from the Lane County Problem Solving in Mathematics Project (Dale Seymour Publications, 1983) to introduce each strategy. The Lane County Project from the University of Oregon presents lessons to specifically teach the problem-solving strategies of guess and check, patterns, systematic listing, tables and graphs, and eliminating possibilities. These lessons give examples for students to discuss and provide ample opportunities for students to practice each strategy. A new strategy is emphasized each week using the class activities and individual assignments. An introductory session from the Lane County materials might go like this:

Teacher: Today we're going to learn a new way to solve problems. The method is called "looking for patterns," and that is exactly what we do. Look at this problem: (Shows 6, 8, 10, __, __, __ on the overhead). You are to decide what numbers go in the blanks. We solve the problem by looking for a pattern in the numbers that are given. Can you find a pattern?

Student 1: They're all even numbers.

Student 2: They go up by 2.

Teacher: Can you decide what numbers go in the blanks?

Student 3: 12, 14, 16

Teacher: Explain the pattern you used.

Student 3: The numbers were increasing by 2 so I added 2 to 10 to get 12, 2 to 12 to get 14, and 2 to 14 to get 16.

Teacher: Good. Let's try another one. (Shows 2, 5, 8, __, __, __). What is the pattern here?

Student 4: I don't think there is a pattern. Nothing is the same.

Teacher: Describe what's happening and maybe you'll find one.

Student 4: The numbers are getting bigger.

Student 5: They go even, odd, even.

Student 6: I know. They're going up by 3.

Teacher: So what numbers should go in the blanks?

Student 7: 11, 14, 17.

Following this exchange, students are assigned worksheets containing similar problems to complete on their own. These exercises are discussed in a later class, with the emphasis on the problem-solving technique used. No grade is assigned.

Our curriculum gives students additional practice with the strategy using a software package. For example, after learning the pattern strategy, students are paired at the computer to use *Muppets: Layer Cake* (Marin Institute CTW, 1981) and *Funhouse Maze* (Sunburst, 1984). For each program, students are asked to record the pattern used and the final number of moves taken to complete the problem. Then they are to try a different pattern to see if the number of moves can be reduced. A discussion follows each software session where students compare patterns and results.

Adding computer software to the Lane County materials gave students a wider vareity of problem-

Table 1 Problem Solving Strategies

Software	Guess & Check	Patterns	Systematic Listing	Tables & Graphs	Eliminating Possibilities	Grade Level	Company
Bagels	X	X	X		X	3-6	MECC
Code Quest #1		X				4-12	Sunburst
Code Quest #2	X		X	X		4-12	Sunburst
Code Quest #3	X		X	X		4-12	Sunburst
Code Quest #4		X				4-12	Sunburst
Code Quest #5	X	X		X		4-12	Sunburst
Code Quest #6	X			X		4-12	Sunburst
Diagonals	X	X			X	5-9	MECC
Factory		X		X	X	5-9	Sunburst
Funhouse Maze	X	X	X	X		4-12	Sunburst
Gears	X	X		X	X	6-12	Sunburst
Gertrude's Puzzles		X				3-6	Learning Co.
Hurkle	X			X		2-6	MECC
Iggy's Gnees		X				K-6	Sunburst
Incredible Laboratory	X	X		X		3-12	Sunburst
King's Rule		X				6-12	Sunburst
Lemonade Stand		X	X	X		3-6	MECC
Muppets: Layer Cake	X	X			X	K-6	CTW
Muppets: Mix & Match	X		X	X		K-6	CTW
Odd One Out		X				K-6	Sunburst
Pinball Construction		X	X	X	X	5-12	Elect. Arts
Pooling Around		X				6-9	MECC
Puzzletanks	X	X	X	X		3-8	Sunburst
Squares	X		X		X	5-9	MECC
Super Factory		X			X	5-12	Sunburst
Taxman	X	X				5-6	MECC
Teaser's by Tobbs	X					4-12	Sunburst
Teddy's Playground		X				K-4	Sunburst
Thinking with Ink	X				X	5-9	MECC
Whatsit Company				X	X	6-12	Sunburst

solving practice. Students were able to see how the strategies could be applied in problem-solving situations other than paper and pencil. The computer also allowed teachers to address individual differences in learning. The components of visual, cooperative, and, in some instances, spatial learning were added. Students who had difficulty relating to paper and pencil activities were helped by the computer work. The motivational effect of the computer was also apparent. Students eagerly anticipated the computer problem-solving experiences, and they began communicating about problems. The students discussed problems, planned strategies, and shared knowledge with each other—as well as with the teacher. The teachers also enjoyed using the computer as an educational tool and indicated a willingness to continue using it following the study. According to the teachers, students demonstrated an increased confidence in problem solving as the project progressed. Whether due to the computer activities or the Lane County materials, students were more willing to try the problems on their own and required less help from the teachers. Students were able to distinguish and use the problem-solving strategies both on the paper and pencil activities and on the computer activities.

To choose software for each strategy, we first reviewed the suggested curricula from the companies. We soon found, however, that we could extend the use of many of the programs. Our curriculum for the strategy of systematic listing shows how versatile many programs can be. As with the pattern strategy, students are given specific instructions regarding systematic listing and worksheets for practice prior to their computer practice. The following activities are given for Muppets: *Mix and Match* (Marin Institute CTW, 1981) and *Bagels* (MECC, 1979).

Day One

1. Load *Bagels*. Choose to work with a four digit number.
2. Use guess and check to find the number.
3. How many tries did it take?
4. Use systematic listing to find the number.
5. How many tries did it take?
6. Try several other numbers recording the number of tries for each.
7. Discuss which strategy seems to be be for *Bagels*.

Day Two

1. Load *Muppets: Mix and Match*
   a. Create two characters recording the head, body, and feet chosen to make each. Also record the name given each creature.
   b. Share your results with the other students
2. Challenge
   a. Create as many different characters as you can in ten minutes. Record the names.
   b. Make a list of the characters the class found. How many were there? How many possible characters are there?

Day Three

1. In your notebook, develop a systematic plan to create all of the characters.
2. Share your plan with the other students.
3. Decide how the groups could cooperate to create all of these characters.
4. Go to the computer to complete the project.

Similar activities can be used with *Incredible Laboratory* (Sunburst, 1984) for older students. A sample list of software and suggested strategies they support is given in Table 1.

The conclusions reached in the original study were mainly from teacher observation. However, a more structured statistical study was completed comparing the effectiveness of the paper and pencil activities with the computer activities. There was no significant difference between the two practice techniques, indicating that the computer activities were at least as effective as the paper and pencil activities. This study also investigated whether gender and cognitive learning style (field-independence/field-dependance, as determined by Witkin's Embedded Figures Test) would relate to student outcomes across the two problem-solving practice techniques. Cognitive learning style and the two-way interaction between cognitive learning style and problem-solving practice technique did produce significant results, as did the three-way interaction among cognitive learning style, problem-solving practice technique, and gender. High scorers tended to be male, field-independent students using the computer practice technique. The computer also appeared to affect results of field-dependent students, who showed some (not statistically significant) improvement in problem-solving skills using the computer practice technique.

Educators and curriculum designers need to help students gain confidence in problem solving by giving them practice in solving a variety of problems and by making the students aware of the steps and methods of the process. Educators must also use tools and materials best suited to a student's cognitive learning style. We have determined that when the question is problem solving, one of the answers is computers.

References

Jussel, M. R. (1988). *An investigation of the effects of problem-solving practice techniques, cognitive learning styles, and gender on problem-solving achievement abilities and problem-solving preferences of fifth grade mathematics students.* Unpublished doctoral dissertation, University of Nebraska, Lincoln, NE.

Lane County Mathematics Project. (1983). *Problem Solving in Mathematics.* Palo Alto, CA: Dale Seymour Publications.

LeBlanc, J. F., Proudfit, L., & Putt, I. J. (1980). Teaching problem solving in the elementary school. In S. Krulik & R.E. Reyes, (Eds.), *NCTM 1980 Yearbook: Problem Solving in School Mathematics* (pp. 104-116). Reston, VA: National Council of Teachers of Mathematics.

Truckson, E. B. (1982). *The effects of heuristic teaching and instruction in problem solving on the problem-solving performance, mathematical achievement, and attitudes of junior college arithmetic students.* Unpublished doctoral dissertation, University of Cincinnati, Cincinnati, OH.

Witkin, H.A., Oltman, P.K., Raskin, E., & Karp, S.A. (1971). *A manual for the Embedded Figures Test.* Palo Alto: Consulting Psychologists Press.

Software Suppliers

Electronic Arts
2755 Campus Dr.
San Mateo, CA 94403

The Learning Company
6493 Kaiser Dr.
Fremont, CA 94555

Marin Institute CTW
Apple Computer, Inc.
20525 Mariani Ave.
Cupertino, CA 95014

MECC
3490 Lexington Avenue North
St. Paul, Minnesota 55112

Sunburst Communications, Inc.
39 Washington Ave., Room VF 414
Pleasantville, NY 10570

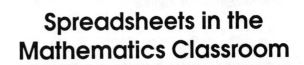

Spreadsheets in the
Mathematics Classroom

The *computer spreadsheet* is a powerful tool for solving certain types of mathematics problems: those involving the storage and mathematical manipulation of large sets of numbers. Until recently, they were used primarily in business and financial situations. But, with the advent of low-cost but powerful spreadsheet programs such as AppleWorks (Apple), PFS: Calc (Apple and IBM), and Microsoft Works (IBM and Macintosh), spreadsheets have been discovered by home users for keeping track of personal finances and by educators for keeping track of grades and as a powerful problem-solving tool for their students.

A computerized spreadsheet may be thought of as a grid of empty rows and columns. The columns are named by letters (A, B, C, etc.) and the rows are named by numbers (1, 2, 3, etc.). A *cell* (a specific location in the grid) is named by the column and row in which it appears (A1, B45, Z245, etc.). See Fig. 2.1. The number of rows and columns available in a spreadsheet depends on two factors: (1) the power of the spreadsheet program itself and (2) the amount of memory available in the computing system on which it is running. Typically, spreadsheets allow more than 100 columns and 1000 rows, with the more powerful of them allowing several times these dimensions. Note that, beyond column Z, columns are named by double letters (AA, AB, AC, etc.).

Depending on the spreadsheet program, there are three or four different kinds of information that may be inserted into a cell. First, there are *labels*—non-numerical or *text* [1] information that is not processed by the spreadsheet when it does computations—intended to identify rows, columns, or other parts of the spreadsheet. Next, there are *numbers*, which may be used in

---

[1] *Text* is simply a sequence of characters—alphabetic, numerical, or special characters like punctuation marks—which is not used in computations. "Name" is text information. So are "43" and "552-43-9921" if the cells containing them have been designated as "text cells" rather than "numeric cells."

spreadsheet computations such as averages, totals, and the like. Third, there are *formulas*, which direct the computer to do an operation on a cell or group of cells (such as finding the sum of the values in a row) and print the results in the cell where the formula is entered. Most spreadsheets also allow you to store dates and times in a cell, including the current date and/or time as kept on the computer's internal clock. The difference between these four types of cell entry is important: a spreadsheet program will do computations on numbers, but not on text; numbers, text, and dates are all sorted differently; and, of course, if you enter a formula as text rather than as a formula the formula itself will be displayed in the cell rather than the results of the computation it does.[2] See Fig. 2.1.

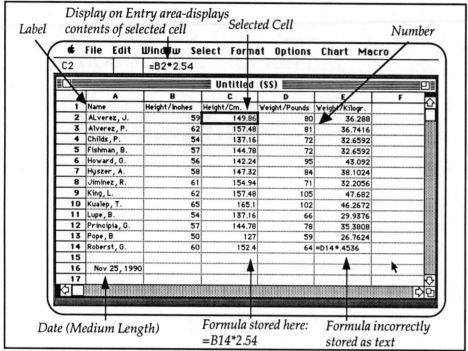

Fig. 2.1: Microsoft Works spreadsheet, Macintosh

Entering information into a spreadsheet is quite simple. First, select the desired cell (use the arrow keys to move to the desired cell, point and click, or use "go-to cell..." from the appropriate menu), then start typing the information you want to put into it. Most spreadsheet programs display the characters you type in a data entry area outside of the spreadsheet (See Fig. 2.1). Then insert this information into the cell when you press the RETURN (or ENTER) key, or press one of the arrow keys or use the mouse to select another cell. With most spreadsheet programs, any sequence of characters that can be evaluated as a number (e.g., 43.5 or -16.829*14) will be stored in number format; any sequence of characters that fit the format for a formula (e.g., =sum(A5:A25), +A5 + A27) will be stored as a formula; any sequence of characters entered in one of the date formats (e.g., Sept. 2, 1990; 9/02/90) will be stored as a date; and anything else (e.g., 541-68-9932, or BOB) will be stored as a label.

Spreadsheet programs allow you to create your own formulas or to use built-in functions. In Microsoft Works, all formulas begin with an = sign. In AppleWorks, formulas begin with a digit, a +, -, or (. To build your own formula, use the appropriate cell names in legal mathematical operations. The operation signs used in formulas are the same as those used in arithmetic and algebra except that * is used in place of x to indicate multiplication and all operations must be explicitly indicated (e.g., to multiply 3 times the sum of the contents of cells C7 and M13, enter =3*(C7+M13) in Microsoft Works, or 3*(C7+M13) in AppleWorks).

---

[2] Most spreadsheet programs have a feature that allows you to print any formulas stored in it rather than the resulting values. The purpose of this feature is to assist the user in finding and correcting errors in a spreadsheet.

Each spreadsheet program has its own peculiar format and spelling for built-in functions and there is some variation in the kinds of functions available. Most spreadsheet programs have functions to find sums, averages, maximum value, minimum value, absolute value, square root, and 25 or so more, including statistical and trigonometric functions. To find the sum of the values in all cells from row 5 through row 100 in column M and print the results in cell M105, you would enter

```
=SUM(M5:M100)    (Microsoft Works)
@SUM(M5...M100)  (AppleWorks)
```

into cell M105. Consult your reference manual for details about using spreadsheet functions.

Another feature of spreadsheet programs is that they allow you to specify the format of the information printed in cells prior to entry or after information has been entered into a cell. For example, labels by default are usually left-justified in cells, whereas numbers are right-justified. You may tell the program, however, to right-justify or center the labels or left-justify numbers in a group of cells should you so desire. And, unless you change the default settings, whole numbers will be displayed with no decimal places showing and repeating decimals will fill the cell with repeating digits. You may change the settings numbers in the cells you select so that numbers are all displayed with two decimal places (most spreadsheet programs automatically round decimals) or as dollar amounts. Consult your reference manual for details.

A powerful feature of most spreadsheets is that they allow you to copy information from one location to another in the spreadsheet. This is especially useful when copying formulas. Formulas may be copied to new locations, either exactly as they were in their original location, or using cell references that are relative to their new location. Suppose, for example, that you were using a spreadsheet as a gradebook as in Fig. 1.6 and you had created a long formula, including weightings for certain scores, to calculate students' total points for a grading period. Rather than retype that formula for each student, you could copy the formula so that in each new row it would use the values in that row rather than from the row where the formula originated. See Fig. 2.2 for an example.

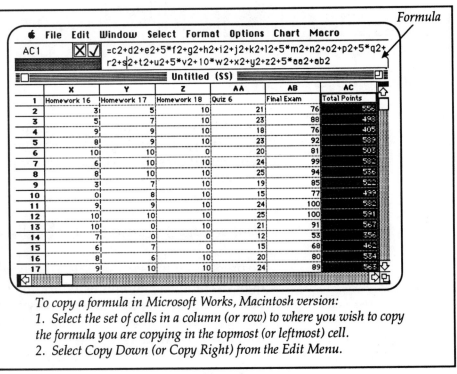

To copy a formula in Microsoft Works, Macintosh version:
1. Select the set of cells in a column (or row) to where you wish to copy the formula you are copying in the topmost (or leftmost) cell.
2. Select Copy Down (or Copy Right) from the Edit Menu.

Fig. 2.2: Spreadsheet showing copying of formulas

Spreadsheet packages such as Microsoft Works and Lotus 1-2-3 have another, powerful feature: they will draw graphs using data in a spreadsheet. Microsoft Works, for example, can draw bar graphs and line graphs of various sorts as well as pie charts. Fig. 2.3 shows a spreadsheet and bar graph done using Microsoft Works. In this activity, students measured their resting heart rate, then exercised. They measured their heart rate immediately after exercise, then after 4 minutes, after 8 minutes, and after 12 minutes. They also recorded the amount of aerobic exercise they did each week. They were asked to discuss the patterns they saw in the data and to speculate what effect regular exercise might have on heart rate. The articles at at the end of this chapter gives a variety of suggestions for using the charting capabilities of spreadsheets like Microsoft Works in the high school classroom.

Name	Resting	Minute 0	Minute 4	Minute 8	Minute 12	AE/Week	
Alverez, J.	76	120	105	82	80	4	
Alvarez, P.	88	133	121	105	99	1	
Childs, P	66	119	105	91	84	5	
Fishman, B.	59	108	92	78	71	4	
Howard, G.	84	105	101	91	90	3	
Hyszer, A.	92	131	117	105	100	0	
Jiminez, R.	83	105	92	88	81	6	
King, L.	91	121	111	104	98	0	
Kualep, T.	76	116	105	91	84	2	
Lupe, B.	79	98	88	81	78	12	
Principia, G	82	99	95	90	88	6	
Pope, B.	88	114	105	97	94	5	
Roberts, G.	64	89	83	79	72	16	
Smith, G.	66	96	88	83	79	2	
Thompson, B.	78	105	103	101	98	1	
Average	78.07	110.60	100.73	91.07	86.40	4.47	

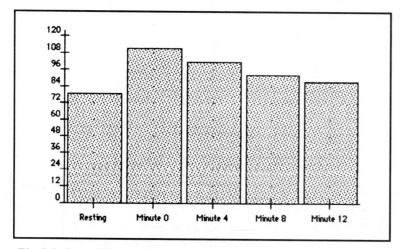

Fig. 2.3: Spreadsheet and bar graph for heart rate data

Student Use of Spreadsheets

The computerized spreadsheet is not a difficult application tool for students to master: entering data and formulas is easy and computation is automatic. And, although it is desirable for each child or groups of two or three children to enter and analyze data that they have gathered in a computer laboratory, the activities discussed below could also be done effectively with one or two computers and accompanying spreadsheets. If you are limited to one or two computers, you may wish to have groups of three to five students gather data and record it on paper, then go to the computer in turn and enter and analyze the data while the rest of the class does other activities. Or, you may wish to enter the data yourself in front of the group (using a large screen or projection monitor), then discuss the results with them.

Introducing Students to Spreadsheets

To prepare students to use spreadsheets, you may wish to spend some time doing traditional graphing activities on a coordinate system (positive quadrant only). Another preparatory activity is to play spreadsheet battleship. This game is similar to the traditional (and commercial) game of battleship: columns are named using letters and rows are named using numbers.

Pupil Activity 2.1: Spreadsheet Battleship

Objective
To practice naming cells on spreadsheets.
Grouping
2 players.
Materials
Each player has a paper grid marked as in Figure 2.4; pencils.
Procedure
A. Out of sight of the other player, each player marks the following on his/her grid (see Figure 2.4):
2 Aircraft Carriers (4 consecutive cells in a column, or a 2 by 2 grid of cells)
2 Battleships (3 consecutive cells in a row)
2 Submarines (a single cell)
B. Players sit facing each other with their playing cards hidden (you may want to place a partition between the players) so that they can see their own card but not their opponent's.
C. Players take turns firing "missiles" at their opponent's ships by naming cells (e.g., "D4"). You may want to have students write the cells they name so that, after a game is over, they can verify that the other person accurately responded to their missiles (marked them correctly).
D. Whenever a player names a cell, his/her opponent responds "hit" or "miss," depending on whether there is a ship at that location and marks each hit with an X. If a hit occurs, the player tells what type of ship was hit. Also, if a player asks his/her opponent how many (or what kinds) of ships are left, the opponent is obliged to tell him/her (*note:* player responds "miss" if the opponent selects a cell formerly occupied by a ship that has been hit).
E. The game is over when one player sinks all of the opponent's ships.

Fig. 2.4: Spreadsheet Battleship playing card

Note: This activity could be easily modified to be played with computers rather than with paper spreadsheets.

After students have learned how to name cells on a spreadsheet, you may wish to have them enter data into a premade template, a predesigned spreadsheet format. A spreadsheet template would include such things as column headings, row labels, formulas, and, possibly, a limited amount of data. The template would be created by the teacher or someone else for a specific problem, then loaded into the children's computers. After the children have gathered the appropriate data, they enter it into the spreadsheet, and discuss the results. Activity 2.2 uses a spreadsheet template for a measurement activity involving area in square centimeters.

Pupil Activity 2.2: Entering Data into a Spreadsheet Template

Objective
Measure metric length, area; entering data into a spreadsheet.
Grouping
Groups of 2 through 4.
Material
Metric rulers; pencil and paper for recording results; spreadsheet template (see Fig. 2.5); computer.
Procedure
1. Select a group of rectangular objects in the classroom. Set up a template as shown in Fig. 2.5, including the appropriate formula in Column D, and print out a copy for each group in the classroom.
2. Have the students measure the length and width of the objects in centimeters and record the results on their paper.
3. After the groups have completed their measurement activities, have them enter their data into the computer and print out the resulting spreadsheet.
4. Discuss the relationship between the area of a rectangle and its dimensions. Perhaps you can lead them to discover the formula for the area of rectangles.

	A	B	C	D
1		Length	Width	Area
2	Table Top			
3	Classroom Floor			
4	Book Cover			
5	Sheet of Paper			
6	Window Pane			
Etc.				

Figure 2.5: Spreadsheet template

Pupil Activity 2.3: Entering Data, Labels, Formulas

Objective
Learn to enter data, labels, and formulas into a spreadsheet.
Grouping
2 to 3 students per computer
Materials
Spreadsheet software; computers; data and problem instructions (see below).
Procedure
Teach the students how to load a spreadsheet and to enter data, labels, and formulas. Have them enter and manipulate small sets of data as in the figure below.

Directions
The chart below shows the number of sit-ups done by six boys in Coach
Jones's exercise class.

	A	B	C	D	E	F	G	H
1		Day 1	Day 2	Day 3	Day 4	Day 5		
2	Sammy	25	28	22	34	42		
3	John	45	41	44	45	49		
4	Jaime	37	42	50	55	56		
5	Charley	12	16	17	23	28		
6	Alex	58	63	67	63	70		
7								

Fig. 2.6: Practice data set for spreadsheet

1. Enter the information from the table into the spreadsheet.
2. In column G, put formulas to calculate the total number of sit-ups each
student did during the five days.
3. In column H, put formulas to calculate the average number of sit-ups each
person did during the five days.
4. In row 7, put formulas that calculate the total number of sit-ups done on
Day 1, Day 2, etc.

Data-Gathering Activities

When you are ready to have students go through the complete procedure of
gathering data, entering it into a spreadsheet, and manipulating the data,
you will need to set the (mathematical/problem-solving) stage for the
experience. For example, if you were doing an integrated math/science
activity involving the growth of plants in different lighting and/or watering
conditions, appropriate instruction on controlling variables in experiments,
statistical concepts, measurement, and the like, would precede or accompany the data-gathering and analysis activities. The following are some
sample data-gathering/spreadsheet activities.

Pupil Activity 2.4: Personal Data

Objectives
Design questionnaire, gather data, evaluate with a spreadsheet, evaluate
results.
Grouping
Entire class.
Materials
Paper, pencils; computer; spreadsheet.
Procedure
Have the students develop a questionnaire with about 15 or so questions
about their classmates which can be answered numerically. Discuss with the
students the types of questions that are relevant (items that give a single,

numerical value), and those that are irrelevant (questions with non-numerical answers, such as "Who is your favorite rock star?" or potentially sensitive or embarrassing items, such as "What is your weight?"). Review and edit the final questionnaire before it is distributed. Next, have the students distribute the questionnaire to the appropriate group. Set up the spreadsheet with appropriate labels and enter the data. Next, enter the formulas to calculate averages, ranges, or other desired statistics. Finally, when the data have been entered and analyzed, discuss the results with the group. For example, "What is the average number of hours of TV that students in this class watch per week?" "Who watches less than the average amount of TV a week?"

Pupil Activity 2.5: Class Party Shopping Comparisons

Objective
Research product prices, enter data, state conclusions about product pricing.
Grouping
Groups of four or larger.
Materials
Dittoed shopping lists, pencils; computers spreadsheets.
Procedure
Have the students develop a shopping list for a class party consisting of about 20 items (e.g., five packages of Kool Aid, three 16 oz. packages of marshmallows, three 16 oz. boxes of graham crackers, and so on.). Be sure to include the brand name and the size of the package so that the prices can be compared. Also, make sure that you use nationally known brands (e.g., Kool Aid) so that the items will be found in all the stores. Next, take groups of five or so children, with the help of parents, to five or six local supermarkets or grocery stores and have them record the price of each item in that store. You may wish to contact the supermarkets in advance to let them know that you are coming. Set up and enter the data into a spreadsheet. Have the spreadsheet total the amount for each supermarket and discuss the results. After deciding which supermarket has the best prices, purchase the items and have a party!

Pupil Activity 2.6: The Cheapest Way from LA to New York
Objective
Gather data, enter and evaluate on spreadsheet, draw conclusions.
Grouping
Groups of 4 or 5 students
Materials
Travel information—flight prices, hotel prices, etc.; restaurant menus; calculators; computers, spreadsheets; pencil, paper.
Procedure
This activity requires that students use a variety of resources—travel agencies, motel/hotel guides, menus or price lists from restaurants, department store catalogs, on-line services, as well as calculators and computers. Start off by discussing which mode of transportation the students think will be the cheapest for traveling from New York to Los Angeles: airline, train, bus, automobile, motorcycle, or bicycle (and/or whatever other modes you

decide on). Discuss the different kinds of expenses that would occur on such a trip (fuel, food, lodging, fairs, etc.) for all modes of transportation. Agree on how nights will be spent, how meals will be obtained. For example, will you eat in restaurants or cook you own meals? If it is decided that the "travelers" will cook their own meals, include the price of camping equipment. Make the necessary resources available, and have students gather and enter the data into the spreadsheet where the different kinds of transportation are listed in the first row as column headings and the row labels are the different kinds of expenses (see Fig. 2.7). After the data have been typed, enter formulas for calculating the sum of each column. If you have a spreadsheet such as Twin, Lotus 1-2-3, AppleWorks GS, or Microsoft Works, or if you have one of the AppleWorks add-ons which draws bar graphs from data, draw a graph of the resulting totals. Discuss the results.

	A	B	C	D	E	F	G	H
1		Air	Train	Bus	Car	Mcycle	Bcycle	Foot
2	N. Days							
3	Ticket							
4	Food							
5	Fuel							
6	Lodging							
7	Repairs							
8	Clothing							
9	Equipment							
10								
11	Total							

Fig. 2.7: Spreadsheet for trip expense comparisons

The article by Edwards, Bitter, and Hatfield in Chapter 3 presents a set of problem-solving activities using spreadsheets and data base managers with data about passengers aboard the ship *Oregon*. This ship carried immigrants to the United States in the 1840s. The spreadsheet activities involve traditional English/American units of measurement, the cost of foodstuffs in the 1800s, and managing budgets.

Mathematical Patterns

The use of spreadsheets in mathematics is not limited to the analysis in data-gathering activities. There are many other uses, including the discovery of general mathematical principles and mathematical patterns. The spreadsheet is an ideal tool for exploring number patterns because of its ability to do quick calculations. Another important feature is the ability to copy formulas to new locations so that they use the values of the row or column in the new location (relative values). For example, if the formula in cell B3 was +B2*2 and the value in B2 was 2, you could, with a few keystrokes, copy the formula in B3 to the next 1000 cells so that each cell would be the double of the previous cell. (See Figure 2.8.)

One type of activity using a spreadsheet is for the teacher to produce a number pattern. It is given to the student on a ditto sheet or on the chalkboard, or already entered into the spreadsheet (which the student loads and manipulates). The student's problem is (1) to continue the pattern for several more cells and (2) to reproduce the pattern by finding, entering, and

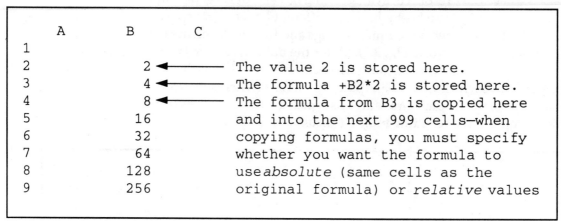

Fig. 2.8: Copying formulas in a spreadsheet

replicating the appropriate formula. When the student has found an answer, he/she prints out the resulting spreadsheet showing the formulas of the figure.

Pupil Activity 2.7: Number Pattern 1
Objective
Continue pattern, find and enter formula for pattern.
Grouping
Individual students or groups of 2 or 3.
Materials
Computers, spreadsheets.
Procedure
Give students the spreadsheet shown in Fig. 2.9, containing instructions, but without the formulas and numbers in cells A15 to A17 at the right.

Pupil Activity 2.8: More Complex Number Patterns
Objective
Continue number series, enter formula.
Grouping
Individuals, groups of 2 or 3.
Materials
Computer, spreadsheet.
Procedure
Give student the spreadsheet shown in Fig. 2.10 (containing instructions).

Pupil Activity 2.9: Advanced Problem
Directions
Enter the following number sequence into a column in the spreadsheet: 1 1 2 3 5 8 13 21 34 45. After the 10th number, enter a formula that causes the pattern to continue for 25 more numbers.

```
              A              B                A              B
1  Directions: Enter the next    Directions: Enter the next 3
2  3 numbers in the pattern in    numbers in the pattern in Column
3  Column A.  In Column B, enter  A.  In Column B, enter the formula
4  the formula that produces the  that produces the pattern.
5  pattern.
6
7     Pattern      Formula         Pattern        Formula
8        5                            5              5
9       10                           10            +B8+5
10      15                           15            +B9+5
11      20                           20            +B10+5
12      25                           25            +B11+5
13      30                           30            +B12+5
14      35                           35            +B13+5
15                                   40            +B14+5
16                                   45            +B15+5
17                                   50            +B16+5
   ( ◄── Original Spreadsheet ──► )  (Printout with   (Printout showing
                                      numerical ans.)     formulas)
```

Fig. 2.9: Patterns on spreadsheets

```
           A          B          C          D          E          F
1  Directions: For each of the patterns in this spreadsheet,
2  enter the next 3 numbers.  In the next column, enter the formula
3  that produces the pattern.  The formula should manipulate the
4  first number (the numbers in columns A and D) to obtain the
5  second (the numbers in columns B and E).
6
7  Number    Pattern 1  Formula 1   Number    Pattern 2  Formula 2
8     1          5          *           1          2          **
9     2          7                      2          5
10    3          9                      3          10
11    4         11                      4          17
12    5         13                      5          26
13    6         15                      6          37
14
15            * (For the reader's information: 2*A8+3)
16            ** (For the reader's information: (D8*D8)+1  or  +D8^2+1)
```

Fig. 2.10: Spreadsheet patterns

The article in this chapter, written by John C. Russell, gives further examples of the use of spreadsheets in investigating patterns. This article describes the use of a computerized spreadsheet to simulate experiments in probability, and hence, to help children understand concepts related to probability. Although the article focuses on a specific (and somewhat unusual) brand of spreadsheet, the activities described could easily be done with virtually any modern spreadsheet package. The article written by Janet McDonald is about ways of integrating computerized spreadsheets into the high school mathematics classroom. She discusses how topics in geometry, algebra, and trigonometry may be explored by students with this powerful tool.

Discovering Mathematical Meanings

The spreadsheet is an excellent tool for discovering a variety of mathematical meanings, especially where sequences of numbers are involved.

Pupil Activity 2.10: Associative Property
Objective
Discover the associative property (multiplication over addition).
Grouping
Individual or pairs.
Materials
Spreadsheet with instructions.
Procedure
Give students the spreadsheet containing directions and data below (Fig. 2.11). *Note*: A prerequisite to understanding the associative property is knowing the standard order of operations and the use of parentheses in mathematical sentences.
Directions
Enter the groups of three numbers in column A, B, and C. Add three more sets of numbers. In column D, enter formulas that first add the numbers in columns B and C, then multipy the results by A (e.g., A1*(B1+C1)). In columns E, F, and G, enter the formulas A1*B1 + C1, A1*C1+B1, and A1*B1+A1*C1. Copy these formulas to rows 3 through 9.

	A	B	C	D	E	F	G
				A*(B+C)	A*B+C	A*C+B	A*B+A*C
1	Number 1	Number 2	Number 3				
2	3	4	5				
3	7	3	6				
4	12	10	15				
5	25	32	15				
6	2.7	1.9	7.25				

What pattern do you see? Complete the following sentence. "Multiplying a number by the sum of two numbers gives the same answer as _____."

Fig. 2.11: Spreadsheet for discovering the associative property.

Pupil Activity 2.11: Estimating Quotients
Objective
Estimate quotients of numbers.
Grouping
 Individuals or pairs of students.
Materials
Spreadsheet with dividends and divisors entered as in Fig. 2.12; column D
has a formula for the product of the numbers in columns B and C.
Procedure
Students estimate quotients of the numbers in columns A and B. The
product of the two factors is displayed in column D. By using their
estimates, students improve their estimate until they have obtained
the exact quotient.
Directions
In column C of the spreadsheet below, enter your estimate of the quotient of
the number in column A divided by the number in column B (i.e., the number
you would multiply by the number in column B to get the number in column
A). The product of the number you entered and the number in column B will
appear in column D. Use that number to improve your next estimate. Try
to find the exact quotient by your
sixth try.

In her article at the end of this
chapter, Barbara Dubitsky (1988),
discusses activities similar to the
one just described in Pupil Activity 2.11 for helping middle school
students understand decimal quotients.

Many other mathematical
concepts and patterns could be
discovered or applied and many
other mathematical skills could
be practiced on computer spreadsheets. Data-gathering activities
offer the opportunity to integrate
mathematics with other subjects
and to explore topics of interest to
children and at the same time give them experiences with statistics, graphing, and the use of spreadsheets. The computer spreadsheet is the perfect
medium to accomplish the repeated computations required of this type of
activity.

	A	B	C	D	E
	A	B	C	D	E
1	Dividend	Divisor	Est. Quot.	B x C	
2	1225	25	10*	250**	
3			40*	1000**	
4			47*	1175**	
5			51*	1275**	
6			49*	1225**	
7					
8					
9	578	17	*Sample student input.		
10			** The formula entered into cell		
	Etc.		D2 and copied in this column is +B2*C2		

Fig. 2.12: *Spreadsheet for estimating quotients.*

Teacher Use of Spreadsheets

In the previous chapter, the use of a spreadsheet program as a gradebook—
to keep track of student scores and points, to calculate averages, etc.—was
presented (see Fig. 1.6). Spreadsheet programs are useful to the teacher for
other tasks, such as keeping track of inventory (e.g., materials in a math lab,
books, and the like) or expenditures. Suppose, for example, that you were

given a budget of $400 for purchasing classroom materials during a given year. Your spreadsheet might look like the following:

```
            A              B         C        D         E        F       G
 1 Jenny Jones' classroom materials expenditures
 2 1991 - 1992 academic year
 3 ---------------------------------------------------------------
 4   Item Name          Vendor    Item #   Ordered   Price   Number Balance
 5                                         9/16/91                   $400.00
 6 Casio HS-7 Calc.     Seymour   RS16855  10/6/91   $4.50     20    $310.00
 7 Meterstick(dozen)    Seymour   RS02742  10/6/91   $11.40     3    $305.80
 8 Pattern Blocks       Seymour   RS02503  10/6/91   $16.50     5    $223.30
 9 Popsicle Sticks      KMart              10/30/91  $2.65      1    $220.65
Etc.
```

Figure 2.13: Using a spreadsheet to keep track of expenditures

There are many other possible uses of spreadsheets by the teacher. You could use one to keep track of students' schedules for using the computers in your classroom. You could use a spreadsheet to set up a more flexible version of your daily plan book. You could set up a spreadsheet to keep track of students' mastery of basic computational facts. You could set up, print out and post spreadsheets containing information about students completion of homework assignments. Such a spreadsheet might contain student names in Column A and assignment names as column headings. Letters would be entered into cells to indicate whether the assignment was satisfactorily completed (X), needed to be redone (R), or missing (M). While it might be just as easy, initially, to set up the traditional homework grid (e.g., with gold stars for each homework assignment) as it would be to set up a homework spreadsheet, the spreadsheet version offers much more flexibility, including the ability to make daily changes and printouts, calculate totals (using the IF function), or to generate graphs.

The computerized spreadsheet is one of the easiest general application tools to learn to use, yet it is one of the most powerful tools available for exploring mathematical topics, for solving problems, and for storing and organizing information. And, unlike many other classroom computing tools, it can be used effectively in the regular classroom setting even if you have only one or two computers available to you.

Bibliography

Arganbright, D.E. 1984. Mathematical applications of an electronic spreadsheet. In V. P. Hansen, Ed. *Computers in Mathematics Education: Yearbook of the National Council of Teachers of Mathematics*. Reston, Va.: NCTM, 184-193.

Bright, G.W. 1984. Teaching mathematics with technology: Mathematics and spreadsheets. *Arithmetic Teacher* 36(8, April): 52-53.

Dubitsky, B. 1988. Making division meaningful with a spreadsheet. *Arithmetic Teacher* 36(3, November); 18-21.

Edwards, N. T., Bitter, G. G. and Hatfield, M. M. 1990. Teaching mathematics with technology: Data base and spreadsheet templates with public domain software. *Arithmetic Teacher* 37(8, April): 52-55.

McDonald, J. L. 1988. Integrating spreadsheets into the mathematics classroom. *Mathematics Teacher* 81(8, November): 615-22.

Sullivan, P. 1988. Spreadsheets: Another aid for the mathematics teacher. *Australian Mathematics Teacher* 44(4, December): 27-30.

Sweet, S. 1987. Plan a picnic. *Teaching and Computers* 4(6, May/June): 15-17.

Troutner, J. 1988. Computer literacy: Using spreadsheets in the elementary and middle school math class. *Journal of Computers in Mathematics and Science Teaching* 7(4, Summer): 12-13.

Verderber, N. 1990. Spreadsheets and problem solving with AppleWorks in mathematics teaching. *Journal of Computers in Mathematics and Science Teaching* 9 (3, Spring):45-51.

Wood, J. 1990. Utilizing the spreadsheet and charting capabilities of Microsoft Works in the mathematics classroom. *Journal of Computers in Mathematics and Science Teaching* 9 (3, Spring): 65-71.

Discussion Questions

1. How is a spreadsheet superior to a calculator in doing computations on rows and columns of numbers?
2. What features/concepts related to spreadsheets might children have a hard time comprehending?
3. Suggest some activities involving geometry using a spreadsheet. Describe one of these activities in detail.
4. Suggest an activity for children involving "what-if" projections. Describe one of these activities in detail.

Activities

Activity 2.1

Objective
Use a spreadsheet to solve number pattern problems.
Rationale
One of the most important capabilities in solving problems is the ability to discover and explain patterns. Spreadsheets provide a powerful tool for exploring and discovering number patterns.
Materials
Computer, spreadsheet; number pattern in Pupil Activity 2.9
Advance Preparation
Read this Chapter. Learn how to enter values, labels, and formulas into a spreadsheet. Learn how to copy formulas.
Procedure
Read Pupil activity 2.9, then set up a spreadsheet to solve the number pattern. Enter the first 10 numbers into cells B3 through B12. See if you can enter a formula into cell B13 to continue the pattern. Copy the formula into the next 25 cells. (Note: try entering the first 3 numbers into column C and see if your formula successfully continues the pattern in cells B6 through B13).

Activity 2.2

Objective
Develop a spreadsheet activity for children involving estimation.
Rationale
The ability to estimate is a very important skill in a technological society. The immediate calculation capability of the spreadsheet helps students evaluate the accuracy of their estimates.
Materials
Spreadsheet, computer, pencil, paper.
Advance Preparation
Ability to enter numbers, labels, and formulas into spreadsheet.
Procedure
Develop a spreadsheet activity similar to Pupil Activity 2.11, except, instead of division problems, the spreadsheet should involve estimating the difference of two numbers (e.g., 3456 - 1638). Column A should contain the minuend, column B the subtrahend, column C the student's estimate, and Column D the sum of columns B and C.

Problem Solving with a Spreadsheet

by Karl Hoeffner, Monica Kendall, Cheryl Stellenwerf, Pixie Thames, Patricia Williams

As mathematics education adapts to the needs of a changing society, the use of technology becomes increasingly important. Technology can play a valuable role in developing problem-solving capabilities in students by freeing them from tedious computations. Thus, higher-level thinking skills can be emphasized. An electronic spreadsheet can calculate rapidly, generate data from which patterns can be found, show relationships between two or more variables, and investigate "what if?" questions with ease.

The use of the activities presented in this article requires some prerequisite knowledge of how spreadsheets work. Teachers need to be able to create simple templates. Students need only to be able to enter data into cells, although more advanced students can be expected to create their own templates. For novice spreadsheet users, these skills can easily be learned. Some school districts offer short training classes, and some spreadsheet programs offer built-in tutorials. The value of spreadsheet use in the mathematics classroom is well worth the time invested.

All the activities discussed here have been successfully used in our classrooms. The students have been very successful at learning to use spreadsheets. They have enjoyed solving problems with spreadsheets and have expressed a desire to work with spreadsheets again. The activities described in this article demonstrate the breadth of spreadsheet applications in the mathematics classroom. Activities 1 and 2 illustrate the use of the spreadsheet as a practical problem-solving tool. Activity 3 shows how the spreadsheet can be used to develop a conceptual understanding of relationships between variables. The template figures used as examples are adaptable to any spreadsheet program.

Activity 1: Holiday shopping

Objective: To select gifts and stay within a given budget

Procedure: Introduce the following scenario to your students:

It's time to think about shopping for the holidays. You have saved $120 and are ready to buy gifts for five people. The spreadsheet will help you keep track of your purchases, including the sales tax that you have to pay. Most important, it will help you make

Figure 1

	A	B	C	D	E	F	G	H
1	Name	Gift	Reg. Price	Disc. Rate	Disc.	Sale Price	Tax	Item cost
2					=C2*D2	=C2-E2	=.08*F2	=F2+G2
3					=C3*D3	=C3-E3	=.08*F3	=F3+G3
4					=C4*D4	=C4-E4	=.08*F4	=F4+G4
5					=C5*D5	=C5-E5	=.08*F5	=F5+G5
6					=C6*D6	=C6-E6	=.08*F6	=F6+G6
7							Total cost	=Sum(H2:H6)

selections that will enable you to stay within your budget. Your goal is to spend at least $100 but not more than $120.

Have students bring in catalogs from various stores from which they can do their "shopping." Each student should select a gift for each of five relatives or friends from the catalogs. The students then enter the name of each person, the gift, the regular price, and any possible discount rate as variables in the template (see fig. 1). Then they make the spreadsheet perform all calculations. Be sure to use the correct tax rate in column G of the template. (In Houston, the tax rate is 8%.) Have the students check to see that the total cost falls within the $100-$120 range. See figure 2 for an example of a student's completed work. Rounding within the computer accounts for the discrepancies of a penny in the sales prices.

Extra challenge

Have your students revise their selections to include gifts for eight people (make sure the template is set up to accommodate eight gifts).

Figure 2

	A	B	C	D	E	F	G	H
1	Name	Gift	Reg. Price	Disc. Rate	Disc.	Sale Price	Tax	Item cost
2								
3								
4								
5								
6								
7								

Figure 3

	A	B	C	D	E
1	Budget	$550.00			
2					
3	Items	Cost per item	#Pkgs. needed	Packages	Total cost
4	Cookies (1)	$0.48			=B4*C4
5	Soft drinks (1)	$0.38			=B5*C5
6	Balloons(10)	$2.66		=C6/10	=B6*D6
7	Plates (20)	$2.38		=C7/20	=B7*D7
8	Napkins (25)	$2.00		=C8/25	=B8*D8
9	Cups (8)	$0.89		=C9/8	=B9*D9
10	Party Favors (7)	$5.50		=C10/7	=B10*D10
11					
12				Total cost	=Sum(E4:E10)
13				Money left	=B1-E12

Figure 4

	A	B	C	D	E
1	Budget	$550.00			
2					
3	Items	Cost per item	#Pkgs. needed	Packages	Total cost
4	Cookies (1)	$0.48			=B4*C4
5	Soft drinks (1)	$0.38			=B5*C5
6	Balloons(10)	$2.66		=C6/10	=B6*D6
7	Plates (20)	$2.38		=C7/20	=B7*D7
8	Napkins (25)	$2.00		=C8/25	=B8*D8
9	Cups (8)	$0.89		=C9/8	=B9*D9
10	Party Favors (7)	$5.50		=C10/7	=B10*D10
11					
12				Total cost	=Sum(E4:E10)
13				Money left	=B1-E12

Have a contest with your students to see who can come closest to $120 without exceeding that amount.

Activity 2: Planning a party

Objective: To purchase items for a party and stay within a given budget

Procedure: Introduce the following scenario to your students:

A student committee has been given the task of planning a school party; $550.00 has been raised for the event, and 150 students are expected to attend. Refreshments, decorations, and party favors must be purchased. Purchases should be made so that each guest can have at least one party favor and as many cookies and drinks as possible. Each new drink will require a new cup, and each new cookie will require a new napkin and a new plate. In addition, 200 balloons will be needed for decoration. This is the shopping list:

> Cookies: $0.48/each
> Soft drinks: $0.38/each
> Balloons: $2.66/package of 10
> Plates: $2.38/package of 20
> Napkins: $2.00/package of 25
> Cups: $0.89/package of 8
> Party favors: $5.50/package of 7

Using the spreadsheet template (see fig. 3),have students enter data into column C to determine the number needed for each item. After entering the data, have the spreadsheet perform the "total cost" calculations to make sure that the budget has not been exceeded. Remind students that they cannot buy part of a package. A sample of a student's completed work is presented in figure 4. Notice that 304 cups and 154 party favors had to be purchased, rather than 300 and 150, respectively, since only complete packages could be purchased. Ask students to answer the following questions :

1. How many cookies, soft drinks, and party favors can each guest have?

2. How much money is left over?

Extra challenge

• If 350 students come to the party, will there be money left over if each has 1 cookie, 1 soft drink, and 1 party favor? If yes, how much money will be left? If no, how much additional money is needed?

• How many students can participate if each has only 1 cookie, 1 soft drink, and 1 party favor?

• If each student reuses his or her cup, napkin, and plate (1 per person) and has 3 cookies, 2 soft drinks, and 1 party favor, how many students can be accommodated?

• If tickets are sold to cover all the party expenses, how much must be charged for a ticket if 130 students attend and each has 4 cookies, 2 soft drinks, and 2 party favors?

Activity 3: Interesting patterns

Objective: To investigate relationships between variables in the simple-interest formula

Procedure: Hold a discussion with the students about borrowing money from a bank and familiarize them with the elements of the simple-interest formula. Have students use the template in figure 5 to investigate the following problem:

> Pat can afford monthly payments of $325 on a used car. How much principal can be borrowed over 3 years at annual interest rates of 5%, 10%, and 15% while keeping the monthly payments at exactly $325?

A guess-and-check strategy for entering the principal should be used by the students. They should have the spreadsheet calculate the monthly payment. Then they should continue refining their guesses until the target monthly payment is achieved for each interest rate.

After the first three rows are completed, encourage students to analyze the data for possible patterns. For example, it appears that when the interest rate is increased by 5 percent, the principal decreases by about $1000 (see fig. 6). Instead of randomly making the initial guess, students can use this information to estimate the principal for the higher interest rates in the template (rows 5, 6, and 7). More guessing and checking will be needed to obtain the exact answers.

Students should quickly discover that a decrease in principal of $1000 is too large, as the interest rate continues to increase. They will need to refine their strategy further to compensate for this factor.

Extra challenge: Use the spreadsheet data to graph the relationship between the interest rate and the principal. Then use the graph to predict the principal for interest rates of 50 percent, 75 percent, and 100 percent. (We know that these are not "realistic" rates, but the graph can serve as a basis for an interesting discussion.)

Figure 5

	A	B	C	D	E
1	Interest	Principal	Int. rate	Time	Mthly. pmt.
2	=B2*C2*D2		.05	3	=(A2+B2)/(D2*12)
3	=B3*C3*D3		=C2+.05	3	=(A3+B3)/(D3*12)
4	=B4*C4*D4		=C3+.05	3	=(A4+B4)/(D4*12)
5	=B5*C5*D5		=C4+.05	3	=(A5+B5)/(D5*12)
6	=B6*C6*D6		=C5+.05	3	=(A6+B6)/(D6*12)
7	=B7*C7*D7		=C6+.05	3	=(A7+B7)/(D7*12)

Figure 6

	A	B	C	D	E
1	Interest	Principal	Int. rate	Time	Mthly. pmt.
2	$1 526.10	$10 174.00	.05	3	$325.00
3	$2 700.00	$ 9 000.00	.10	3	$325.00
4	$3 631.05	$ 8 069.00	.15	3	$ 0.00
5	$ 0.00		.20	3	$ 0.00
6	$ 0.00		.25	3	$ 0.00
7	$ 0.00		.30	3	$ 0.00

Making Division Meaningful with a Spreadsheet

by Barbara Dubitsky

Long division is such a complicated procedure that it takes most students many years to learn all the variations. Even with a great deal of practice, many get so bogged down in the manipulation of numbers that they lose sight of what they were trying to solve. Students often come up with absurd answers and don't question their correctness.

A technique that uses an electronic spreadsheet can help students understand long division. The electronic spreadsheet can free students from the tedium of number manipulation and allow them to concentrate on the problem and find meaningful answers. Then they can solve division problems with pencil and paper with more understanding.

What are electronic spreadsheets? They are computer programs invented for the business world to take the place of the large paper spreadsheets that were used for financial planning. The interactive nature of the electronic spreadsheet makes it easy for people to ask what-if questions about business problems. They can also be used in schools to allow students to ask what-if questions about mathematics.

Figure 1 A Portion of the Spreadsheet

C3

	A	B	C	D	E
1					
2					
3			/ / / / /		
4					
5					
6					
7					
8					
9					
10					
11					
12					

The electronic spreadsheet, designed primarily for the manipulation of sets of numbers, functions like a huge piece of paper. The smallest unit of the spreadsheet's construction is the cell. Each cell is named with a column letter and row number. In figure 1, the cursor (indicated by ///// in the figure and by a white rectangle on the computer screen) is in cell C3. This information is shown in the top left corner of the screen. The user can insert material into the spreadsheet by placing the cursor at a given cell and typing in the appropriate information in that cell.

Division word problems can be typed onto the spreadsheet and presented to students. Below the problem, a column is provided for the student to make a series of guesses at the answer. With each guess the student gets feedback about the closeness of that guess to the answer until a solution is found. Since it is so easy to change numbers on a spreadsheet, one can ask the student to solve many versions of the same problem. Students can find patterns and discover their own techniques for solution.

Although we continually talk about guessing, the students with whom we worked did not actually guess; they were always thinking about the reasonableness of their number choices. We had hoped that they would take more chances and actually do some guessing, but they didn't think that was a proper way to do mathematics. (We know that mathematicians are always making guesses and then figuring out how to get closer to the solution.) As a general rule, we did not let students use pencil and paper to do these problems because that approach usually defeated the purpose of letting them think more freely about the mathematics they were doing.

An Example with Powers of Ten

We were working with seventh graders who were having difficulty with mathematics. Generalizing the pattern of dividing by 10, 100, and 1000 was a concept most of them had never understood. The following problem had been typed onto the spreadsheet before the session:

> 320 bottles of soda are shared equally among 10 classes. How many bottles of soda does each class get?

Table 1 illustrates the way the problem looked on the spreadsheet with columns underneath for the student to enter estimates of the number of bottles of soda each class should receive. The columns were set up on the spreadsheet so that when the student entered a number in the first column ("Number of Bottles to Each Class"), the total given out appeared in the next column ("Total Given Out") and the difference between this amount and 320 appeared in the final column. The student typed successive "guesses" under each other in the first column. This material was introduced as a game in which the student typed in guesses at the solution. The game was won when a zero appeared in the last column. The students were asked not to use paper and pencil to derive solutions.

Two girls who were seated at one of the computers tackled the first problem together. When they had solved it, they were asked to substitute 850 for 320 and erase the previous set of trial solutions. Different numbers were substituted for the number of bottles of soda until they saw a pattern, which they articulated. Then they were asked to change the number of classes to 100 and the number of bottles to 7800. They solved this variation easily and were eager for other problems. Soon they came rushing to the teachers to say that they could do any problems like this and expounded on why they were so easy. Their regular mathematics teacher was present and assured us that she had previously taught these girls the same generalizations, but they had never understood the concept as they seemed to now.

Table 1 A Problem on a Spreadsheet

	A	B	C	D	E	F
1	320	bottles of soda are shared equally among				
2	10	classes. How many bottles of soda does each				
3		class get?				
4						
5		Number of			Total		
6		Bottles Each*			Given out		320
7						
8		20		200		-120
9		40		400		80
10		38		380		60
11		35		350		30
12		32		320		0

*Student's guesses

When working with whole classes, it is useful to prepare sheets like the one in table 2, giving the student a series of numbers to place in the problem. It furnishes space for a correct solution and allows the student to see the emerging pattern.

On another occasion, we were working with sixth graders who had had experience with division but whose conventional methods prevented them from knowing when a solution was reasonable. They had never considered the use of estimation in division. They began working with the problem illustrated in table 3.

The students were asked not to use pencil and paper but to guess at answers. Often, their initial guesses were far from correct. All the students understood when their guesses were too big or too small, but they couldn't easily narrow their solutions to the correct one. We found that it wasn't helpful to ask such leading questions as "Why can't you give a thousand to each school?" or "How many hundreds can you give to each?" Instead, we allowed the students to find their own ways of solving the problems. Sometimes we helped a student focus by asking if his or her last guess was too big or too small. We gave children many versions of the same problem and, with guidance to assure that solutions would be whole numbers, allowed students to enter numbers of their own choosing. When they tired of this problem, we went on to others that required the same skills. It took experience with many such problems over several sessions before a number of students were able to invent their own estimation techniques. Eventually, all the students came up with their own ways of estimating solutions. Most often, they used the techniques of looking at multiples of ten that we found we could not "teach" them.

Table 2
A Sample Table for Students to Record their Answers

Number of bottles	Number of classes	Number of bottles each
320	10	
850	10	
3590	10	
6700	10	
7800	100	
6700	100	
83000	1000	
94000	100	

Table 3 Student's Guesses without Using Paper and Pencil

	A	B	C	D	E	F
1	744 computers were shared equally among 8 schools.					
2	How many computers did each school receive?					
3						
4	Number given to		Total		Number	
5	each school		Given out		left over	
6						
7	5		40		704	
8	15		120		624	
9	200		1600		-856	
10	80		640		104	
11	92		736		8	
12	100		800		-56	
13	95		760		-16	
14	94		752		-8	
15	93		744		0	

Table 4 An Example Using Decimal Fractions

	A	B	C	D	E	F
1	10 children shared 25 pieces of bubble gum so					
2	that each child got the same amount. How many pieces					
3	did each child get?					
4						
5						
6	Number of		Total		Number	
7	pieces each		given out		left over	
8						
9	2		20		5	
10	3		30		-5	
11	2.5		25		0	

An Example with Decimals

Sixth graders who had learned about decimal fractions but had never used them in division were given the problem illustrated in table 4. After a few tries at a solution, we heard complaints from around the room: "This computer is broken! I tried 2 and I tried 3 and nothing is working." We responded by asking if they had ever heard of decimals. The responses were big smiles and fingers back on computer keyboards. They quickly taught themselves how to do these problems and were able to discuss the patterns they saw and what rules applied to this type of division problem. They were then ready for the more complex problems introduced in their regular mathematics class. The worksheet used with the problem in table 4 is shown in table 5.

Table 5
A Table for Results of Problems Like Those in Table 4

Number of children	Number of pieces of bubble gum	Number of pieces each
10	25	
10	43	
10	78	
10	364	
100	576	
100	3879	
1000	6734	
100	87	
100	3	
100	7	

Conclusion

We used an electronic spreadsheet in this way with students from the fourth through the seventh grades. Some of them were poor mathematics students, others were quite accomplished. All were able to use the spreadsheet in the way described. They quickly learned to enter, change, and delete numbers and to move from problem to problem. They not only learned about division but began to understand more about numbers in general. They were also learning to form and trust their own theories about mathematics. And we also learned—to be quiet, to watch and listen to the students, and to know more about how they thought about mathematics. We found that, given the right tools, encouragement, and admiration for unique ways of thinking, the students were able to find their own techniques and solutions. They had learned fine problem-solving skills.

Table 6 Formulas Underlying Table 1

At cell	Enter	When asked for relative or no change, respond with
F6	+A1	
D8	+A8*A2	
	/R	
	D8...D8-D9...D12*	(R,N)*
F8	+D8-A1	
	/R F8...F8:F9...F12**	(R,N)*

In every other case type the word or number indicated in table 1.

* In VISICALC, this copies the formula in D8 to D9 to D12 (using relative locations).
** This copies the formula in F8 to F9 to F12 (using relative locations)

Special formulas must be entered on the spreadsheet to calculate the effect of the student's guess. Table 6 gives the formulas and method of entering them onto a VisiCalc spreadsheet for the problem illustrated in table 1. The formulas would be similar for other spreadsheets.

Probability Modeling with a Spreadsheet

by
John C. Russell
edited by Jerry Johnson

Probability filters into the mathematics education of elementary schoolers when, by some accident, the school year outlasts the drill sheets in long division. Or, an inquisitive student may stumble upon the last chapter in the text while the teacher is psyching the class to give Roman numerals their proper respect. In high school, once students have acquired a considerable facility with arcane algebraic maneuvers, one course in probability and statistics may be available in time for departure to college or a career.

The National Council of Teachers of Mathematics claims it is important for students to enter into arenas in which they must consider likelihoods and interpret their chances of success and failure in the future as a result of their observations of the past—in short, deal knowledgeably with data and the underlying probability questions. *USA Today is* a daily reminder, through its reporting of numbers and graphs, of the need for young citizens to be exposed early and significantly to the study of probability.

Two approaches to probability study are common. After-the-fact probability has students estimate the chance of a future occurrence based on what has happened over time. Often the inherent difficulty in this method lies in the inconvenience of collecting data which are both valid and ample. Before-the-fact probability does not rely on the gathering of elusive data, but rather (for all but the simplest of situations) the application of mathematical mechanics available only to students of intermediate algebra and beyond. Such calculating ability obviously is beyond the typical elementary student.

I believe, as many do, that probability is an important study, one that deserves the attention of students at an earlier age than has been the tradition.

Probability concepts must be applied not only in mathematics classes, but in a variety of other contexts. Here are some specific ways to do so.

Consider using a spreadsheet to manufacture probability models. *BetterWorking Spreadsheet** (from Spinnaker Software) accesses the Apple's random number generator, permitting the user to induce "chance" into any cell of the sheet. (Commodore 128 version is also available.) Coins can be flipped and dice can be rolled electronically with relative ease. Hundreds of trials can be simulated without handling a single die or watching a single coin disappear across the classroom floor.

Figure 1

```
                         DICE THROW

A
B
C   Tap ! and see dice rolled in row F with totals and percents below.
D
E
F   2 DICE ->              2            4
G
H              TOTAL 2'S            2     3.23
I              TOTAL 3'S            2            3.23
J              TOTAL 4'S            7           11.29
K              TOTAL 5'S            7           11.29
L              TOTAL 6'S            7           11.29
M              TOTAL 7'S            9           14.52
N              TOTAL 8'S           10           16.13
O              TOTAL 9'S            7           11.29
P              TOTAL 10'S           6            9.68
Q              TOTAL 11'S           3            4.84
R              TOTAL 12'S           2            3.23
S
T              #  THROWS           62

           0            1            2            3
```

As an example, let's look at one of the inevitable early experiments, the rolling of two dice and the tallying of the sums, two through 12.

This model "rolls" the dice on row F. The function in cells F1 and F2 is INT(6*RND (1))+1**, an expression familiar to BASIC programmers. In column 2 the computer increments the total of the appropriate sum. To illustrate, cell K2 contains the expression IF

(F1+F2=5,K2+1,K2), calling upon the total of fives to bump up by one if the two dice on row F do, indeed, display a total of five. Otherwise, K2 remains unchanged for that roll. Column 3 maintains cumulative percentages for each total while cell T2 records the number of throws that have taken place.

I don't suggest that the electronic model be used exclusively in this study. Surely it is important that students handle the dice some, that they examine the faces, that they tabulate their data—that they gain a feel for the real thing. Transfer to the spreadsheet will become appropriate when rolling dice and recording numbers become tedious chores which detract from the study.

When in the manual mode, the student must tap the ! key to to calculate the sheet, and *Better Working Spreadsheet* has an annoying habit of beeping with each calculation of the sheet.*** There is an automatic display feature, though, with which a student can set the sheet to work and have it continually generate data at any specified rate, and this mode has the added benefit of being quiet during the sheet's work. Many data points can be collected over the lunch hour, barring a power outage.

If dice can be rolled, so too can coins be flipped, darts thrown and colored balls drawn from an urn—all electronically. While some students may be able to program the spreadsheet to perform these experiments, teacher-written templates allow all students to perform them.

My experiences with young students have shown that spreadsheet modeling of probability experiments allows mathematical problems a downward mobility previously impossible. For instance, consider the question, "Of 10 random people who walk into the same room, what is the probability that there is at least one pair of matching birthdays?" Elementary school children would not routinely handle this question because of their lack of prerequisite mathematics. Indeed, elementary children tend to think the likelihood of a match is much greater than in reality. All children have a friend whose birthday happens to match Uncle Charlie's, and their younger sister was born the same day as Michael Jackson. So a classroom opinion is likely to fail to recognize the notion of 10 *random* people.

I submit that using the model below, while it will not bring elementary students any closer to an a *prior* formula for figuring the probability, will give them a much improved feeling for the situation.

The formula in cells J1 through S1 is INT (365*RND(1))+1**** (yes, ignoring leap years!). Column 2's strategy is to register a 1 only if the birthday to the left matches a birthday below. Cell P2, for example, contains IF (P1=Q1 OR P1=R1 OR P1=S1, P2+1, P2). If students will take the time to get in tune with the changing numbers in column 1, they will come to feel that matching birthdays are not nearly as likely as their experience might have led them to believe. Using spreadsheet models, young children can come closer to an understanding of the implications of a probability problem, if not a solution.

Statistics is the close cousin of probability, and my last example gives a suggestion in that direction.

The number of dice being rolled (6) and the number of rolls (10) are purely arbitrary for this example. There are 60 cells generating random dice rolls, but of interest here are the sums, means, maximums and minimums around the edges of the sheet. Through continual manual calculation, students will be impressed with certain consistencies in those statistics. A good teacher can even bring into the young stu-

Figure 2

A			
B		**A Birthday Problem**	
C	Ten (10) 'random' people enter a room. What is the probability		
D	of at least two matching birthdays?		
E			
F			
G		Match w/	Total
H	Person #	Birthday	below?	trials
I				
J	1	81	0	68
K	2	31	1	
L	3	295	0	Total
M	4	104	0	success
N	5	193	0	6
O	6	278	0	
P	7	31	0	Prob'ty
Q	8	126	0	of
R	9	185	0	Success
S	10	17	<BLANK>	.08823529
T				
	0	1	2	3

Figure 3

```
A              Six Dice Statistics
B
C
D
E   Roll #    #1    #2    #3    #4    #5    #6    Sum   Mean
F
G     1       5     1     3     3     3     1     16    2.67
H     2       5     3     5     1     2     3     19    3.17
I     3       3     4     2     1     1     3     14    2.33
J     4       1     5     5     6     4     6     27    4.50
K     5       3     2     5     1     1     1     13    2.17
L     6       6     3     3     1     2     1     16    2.67
M     7       4     4     3     4     3     3     21    3.50
N     8       6     5     2     5     3     6     27    4.50
O     9       3     6     2     5     2     2     20    3.33
P    10       6     5     5     2     5     4     27    4.50
Q   ===============================================================
R   Max>      6     6     5     6     5     6     27    4.50
S   Min>      1     1     2     1     1     1     13    2.17
T   Mean    4.20  3.80  3.50  2.90  2.60  3.00  20.00  3.33

    0      1     2     3     4     5     6     7     8
```

dents' awareness the notions of variance and standard deviation, perhaps using a term such as "wandering factor." There is no need to worry about vocabulary or exact quantities. This model and others similar to it help students see that certain numbers "wander around" more than others, and that being able to predict how much they wander just might be useful.

One last note: Apple's random numbers start at the beginning of a list upon powerup; *BetterWorking Spreadsheet* does not work around that, unfortunately. My students record on disk a file called "THROW AWAY" which we build prior to experimentation. Each student picks a number of random numbers for the spreadsheet to throw away. Certainly it is necessary to impress upon students the importance of throwing away a different number of random numbers at the beginning of each session. And to those purists who would complain that the Apple does not really generate random numbers, they're close enough for the purposes here. After all, classroom dice have nicked corners and coins often carry a tad of Juicyfruit stuck to one side. We do not spend enough time in the study of probability with our students. Mathematical immaturity and data collection problems often have stood in our way. I submit that a library of spreadsheet models in the hands of a good teacher can help.

Software

BetterWorking Spreadsheet, Spinnaker Software Corp., One Kendall Sq., Cambridge, MA 02139; ph. 800/826-0706. Apple version, $59.95; Commodore 128 version, $29.95.

Notes from the editor:

* Most modem spreadsheets such as AppleWorks, Microsoft or Lotus 1-2-3, use letters to identify columns and numbers to identify rows. In BetterWorking Spreadsheet, it is just the opposite. Also note that the column labels on BetterWorking Spreadsheet are at the bottom of the screen and the first column in the spreadsheet is identified by a Ø rather than a1.

** Check the reference materialsfor specifics about functions and formulas on your spreadsheet. In Microsoft Works, the formula is =IND(6*RAND( )). Unfortunately, this activity cannot be done with AppleWorks, since it does not have a random function.

*** Most spreadsheets have a recalculate function. If you are using Microsoft Works, selecting Calculate Now from the Options menu or pressing F9 will cause a new pair of random numbers to be generated and the totals based on these two numbers to be recalculated.

**** In Microsoft Works, the formula is =INT(365*RAND()).

Utilizing the Spreadsheet and Charting Capabilities of Microsoft Works in the Mathematics Classroom
by Judith Body Wood

One area that is being neglected in the mathematics classroom is the use of the computer as a tool. Computers are, and will continue to be, a part of our everyday lives—most probably as tools. Almost no one 'types' anymore. Everyone 'word processes.' As the cost of hardware and software continue to decline, more use is expected of these tools in the contemporary classrooms of today.

Figure 1

	y = 3x	y = 3x + 5
x	y	y
-3	-9	-4
-2	-6	-1
-1	-3	2
0	0	5
1	3	8
2	6	11
3	9	14

The hand-held calculator was the innovation in the early 1970s. Much research was conducted on the usefulness and the appropriateness of these aids (or tools) in the mathematics classroom. Today, a more powerful tool is available—computer hardware and software—that is not being used as a learning device under the control of the student. Learning is an active, not a passive, activity. The more the student is involved in an active learning environment, the more learning that occurs.

The purpose of this article is to provide the reader with selected ideas of how to use 'tool' capabilities in the mathematics classroom and to have learning take place. The application does not have to be calculus—general mathematics students can greatly benefit from being exposed to computer usage.

The *Microsoft Works* spreadsheet with its charting (graphing) capabilities is a very good choice for the mathematics classroom. The integrated package is easy to learn and use. It is networkable with the appropriate site license. The cost of *Works* is not extravagant. It can be purchased for less than $100. Included in the package is a tutorial for learning *Microsoft Works*.

Algebra I

Use the spreadsheet and charting feature of *Works* to teach slope. Have students enter data into the spreadsheet, define a chart, and draw a line graph. On the spreadsheet they can view the amount of change in y for each change in x, thus, understanding the concept of slope. Using the charting capability, the students will be able to view the graph of the line. As the slope changes, they can see the graph of the line change. As the students enter new data for the spreadsheet, the chart is automatically changed. The student can nip back and forth to see the effects of changing the values of x and y and the change in the graph.

Figure 1 shows the spreadsheet data. Figure 2 shows the chart produced from the spreadsheet. As shown in Figure 2, one could extend this discussion to include y-intercept and parallel lines. The usefulness of this tool is almost unlimited. After several applications, the students should grasp the intended concepts.

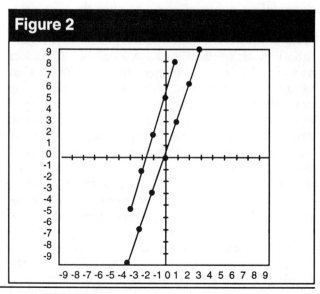

Figure 2

Algebra II

In second year of advanced algebra have students use the Works spreadsheet to calculate values for an exponential function. The spreadsheet data is shown in Figure 3.

Again, use the charting capabilities to draw the graph of the function. The students will easily be able to view the shapes of the curves (as illustrated in Figure 4) of $y = 2x$ or $y = 2-X$, etc. The desired understanding of the concept of exponential functions can be accomplished with little algebraic rigor.

Figure 3

	$y = 2\char94 x$		$y = 2\char94 -x$
x	y	x	y
8	256	8	0.0039063
7	128	7	0.0078125
6	64	6	0.015625
5	32	5	0.03125
4	16	4	0.0625
3	8	3	0.125
2	4	2	0.25
1	2	1	0.5
0	1	0	1
-1	0.5	-1	2
-2	0.25	-2	4
-3	0.125	-3	8
-4	0.0625	-4	16
-5	0.03125	-5	32
-6	0.015625	-6	64
-7	0.0078125	-7	128
-8	0.0039063	-8	256

Figure 4

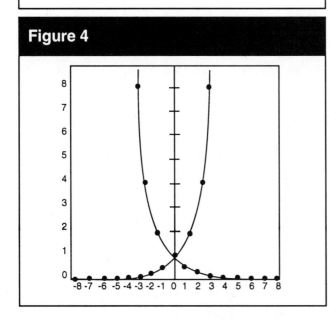

Figure 5

```
HIGH TEMPERATURE FOR THE DAY

Day          Dallas        New York
----------------------------------------
Mon            82             60
Tue            85             54
Wed            80             58
Thur           78             62
Fri            76             68
Sat            75             70
Sun            81             73
```

Thus, the meaning does not become lost in the tediousness of calculation and graphing.

Encourage the students to explore using various exponential functions and other types of functions. They may make some amazing discoveries.

General Mathematics

Perhaps one of the best things that can be done for the General Math student is to make mathematics meaningful. To make mathematics interesting and enjoyable for these students is a great challenge. Imagine the delight of a student that is 'using a computer' in General Math. Have them keep a record of daily temperatures for a week of any two cities. Enter these data into a spreadsheet and have them plot the lines graphs together. Figure 5 shows sample

Figure 6

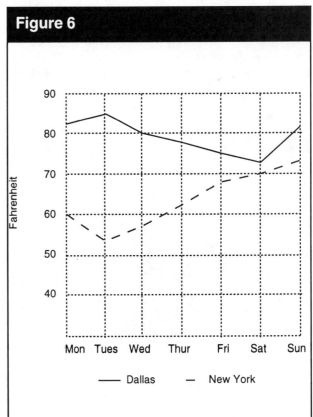

data for two cities, Dallas and New York with the line graphs shown in Figure 6.

Another idea would be to have the students maintain a record of expenditures during a week and draw a pie graph showing percentages spent on each item (Figures 7 and 8). Have.them use the same data to draw a bar graph (Figure 9) and ask them questions in comparing the two graphical representations.

Additionally, one may chose to use graphs produced by the chart just for interpretation. Allow the students to suggest their own ideas for charting.

Figure 7

```
         Expenses  For  The  Week


     Gas            $5.00
     Lunch          $4.00
     Movie          $8.00
     Snacks         $3.00
     Misc.          $5.00
```

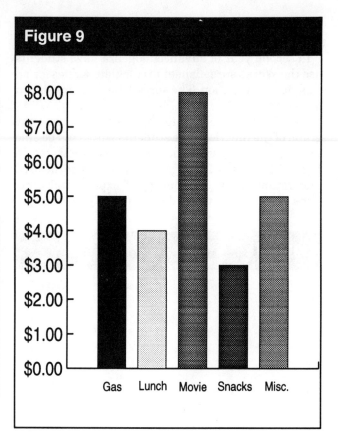

Figure 9

Summary

This article describes just a few ways to use the tool powers of the computer in teaching of mathematics. Further, the students will be learning a useful skill—the use of a spreadsheet. You will find that the students will learn very quickly how to make the computer 'do what they want it to do.' Use your creativity or the creativity of the students to enliven mathematics.

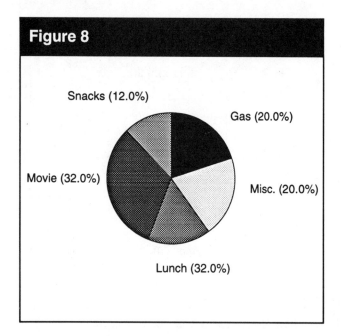

Figure 8

Integrating Spreadsheets into the Mathematics Classroom

by Janet L. McDonald

Spreadsheets have become an integral part of computer literacy and business courses, allowing students to see the power of such utility software and use it to solve problems. But, the spreadsheet can also be an extremely effective tool in the mathematics classroom. There the spreadsheet can be used to help solve many real-world problems and, at the same time, promote students' understanding of important mathematical concepts and principles.

Like the calculator, the spreadsheet not only allows students to experiment with numbers and develop numerical concepts while playing "what if" but also allows teachers to choose more realistic problems without being concerned about extremely difficult calculations. Unlike the calculator, however, the spreadsheet allows students to see a progression of calculations on the screen all at once, thus enabling them to see patterns develop. Students can then make hypotheses based on the data shown and watch on the screen the immediate effect of the changes they make. Thus, in a short period of class time, spreadsheets can offer students a large number of inductive experiences that allow for individualized pacing, more active student involvement, and the potential of the joy of discovering complex mathematical relationships on their own. As in writing or using BASIC programs, students can store on disk and print out any finished or unfinished results for later reference and additional class work, but spreadsheets are often easier to create than programs.

This article describes some techniques for integrating spreadsheets into the mathematics classroom. The examples and discussion will not be directed at any one particular commercial spreadsheet, and no attempt will be made to instruct the reader in the operation and use of spreadsheet software. These examples are appropriate to available software packages.

A variety of approaches can be used to integrate spreadsheets in the classroom, and these approaches will differ depending on the teacher's goals for the lesson, the students' knowledge of spreadsheets, and the level of student involvement desired. For the purposes of this article, it is assumed that the teacher would prepare the entire spreadsheet including all labels, formulas (formulas for the spreadsheets given in this article are listed in the Appendix using a common symbolism), and possibly some initial values prior to class. The teacher would create the table, or template, and save it on disk. Students would merely load the template, enter certain numerical values, and look at the results. In this format, the spreadsheet acts as a simulator, and through a guided discovery approach, the teacher leads students to see general mathematical principles or draw specific mathematical conclusions. The following are examples from several topic areas.

Geometry

Figure 1 shows a template for investigating relationships involving the measures of angles of regular polygons. The spreadsheet would be set up by the teacher with the formulas for the angle measures, the number of sides, and the increment. Students would be required to supply only the values for the number of sides and the increment. The figure shows how the template would appear on the screen after the student entered a starting number of sides of 3 and an increment of 1. A teacher could either use the spreadsheet in a whole-class demonstration-discussion mode using a video projector or give handouts with the sequence of steps and leading questions to be answered in a laboratory environment. This table could be used to lead students to derive the formulas for the angle measures or make generalizations about relative angle sizes and sums. Students, by the way, are continually fascinated with determining the sum of the measures of the interior angles of such regular polygons as the 156-gon. Several variations of this lesson are possible. One would be to add a column to determine the sum of the measures of the exterior angles for each polygon. Another would require students to derive and enter the formulas, compare different-looking but equivalent formulas, or create the entire spreadsheet. Yet another would eliminate the increment and starting-number-of-sides options and be designed to allow students to enter their own individual set of sides in the first column. This final approach does not generate a full table as quickly as that in figure 1, but it allows students to test a greater range of values on one screen and to form their own conjectures one step at a time, testing out each one individually. Figure 2 shows such a template after a student has entered several entries in column A.

Algebra

Functional relationships

Precreated templates can also be used to give students practice in determining functional relationships. Students would be given an incomplete table of values and challenged to determine the relationship by supplying values for x and seeing what corresponding values of y are generated. Figure 3 shows a prepared template after student interaction. Students can check their own conjectures by testing out additional values. If the spreadsheet used has graphics capability, which several spreadsheets do, then with a few simple commands this same table of values can be used to produce a graph of the function as well.

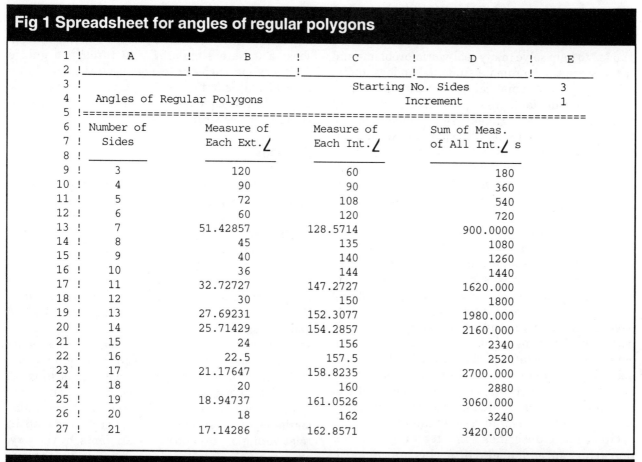

Fig 1 Spreadsheet for angles of regular polygons

	A	B	C	D	E
1					
2					
3			Starting No. Sides		3
4	Angles of Regular Polygons		Increment		1
5					
6	Number of	Measure of	Measure of	Sum of Meas.	
7	Sides	Each Ext. ∠	Each Int. ∠	of All Int. ∠ s	
8					
9	3	120	60	180	
10	4	90	90	360	
11	5	72	108	540	
12	6	60	120	720	
13	7	51.42857	128.5714	900.0000	
14	8	45	135	1080	
15	9	40	140	1260	
16	10	36	144	1440	
17	11	32.72727	147.2727	1620.000	
18	12	30	150	1800	
19	13	27.69231	152.3077	1980.000	
20	14	25.71429	154.2857	2160.000	
21	15	24	156	2340	
22	16	22.5	157.5	2520	
23	17	21.17647	158.8235	2700.000	
24	18	20	160	2880	
25	19	18.94737	161.0526	3060.000	
26	20	18	162	3240	
27	21	17.14286	162.8571	3420.000	

Fig. 2. Sample student choices for sides of polygons

	A	B	C	D	E
1					
2					
3	Angles of Regular Polygons				
4					
5	Number of	Measure of	Measure of	Sum of Meas.	
6	Sides	Each Ext. ∠	Each Int. ∠	of All Int. ∠ s	
7					
8	5	72	108	540	
9	12	30	150	1800	
10	20	18	162	3240	
11	42	8.571429	171.4286	7200.000	
12	100	3.6	176.4	17640	
13	156	2.307692	177.6923	27720.00	
14	1002	.3592814	179.6407	180000.0	
15	50123	.0071823	179.9928	9021780	

Fig. 3 Function Machine

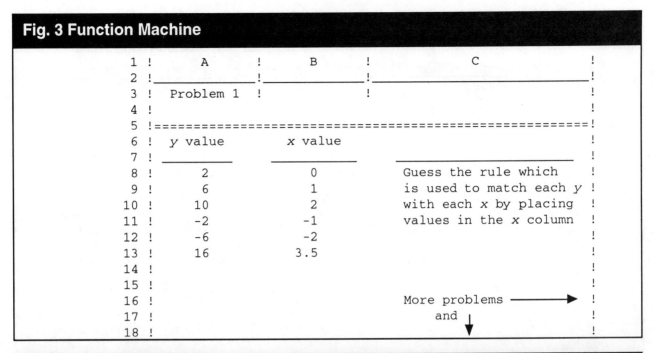

Fig. 4 Hidden coefficients for problems

		BB	BC	BD	BE	BF	BG
99				x^3	x^2	x	k
100	Problem 1					4	2
101	Problem 2				1		-3
102	Problem 3				2	-3	1
103	Problem 4				2	3	
104	Problem 5			1			-2
105	Problem 6			1	1	1	1
106	Problem 7			2		1	

Several special techniques are necessary to make examples like those in figure 3 effective (O'Connor 1985). First, it is necessary to create a window to keep students from going to a cell in the y-column and getting the answer immediately; see your spreadsheet manual for creating windows. As a second line of defense, the formulas that are stored in the y-column should be expressed in terms of variables hidden elsewhere on the spreadsheet. For example, rather than listing 4 * B8 + 2 as the formula in the y-column at cell A8, store it as BD100 * B8 * B8 * B8 + BE100 * B8 * B8 + BF100 * B8 + BG100, where the missing coefficients 4 and 2 are " hidden" in cells BF100 and BG100, as shown in figure 4. Obviously, if a student knows enough about spreadsheets and is determined enough to cheat, a correct solution can be obtained surreptitiously, but most students prefer to determine the answers honestly.

To allow for students working at different rates, spreadsheets like the one shown in figure 3 can also include more problems of the same or similar types placed below or to the side. For example, other problems on this same spreadsheet might include quadratic or cubic relationships in varying degrees of difficulty. The teacher can even identify in the top section of the spreadsheet which problems individual students are to solve. An additional advantage to setting up the template with multiple examples is that if the teacher writes all the functions in terms of general coefficients hidden elsewhere in the spreadsheet, all the problems can be changed by only changing the coefficients rather than by creating whole new templates. Figure 4 shows such a series of coefficients as they might be hidden in a separate portion of the spreadsheet.

Fig. 5 Solving a system of equations

	A	B	C	D	E	F	G	H	
1 !									!
2 !	Equation 1:	5		y =	2	x	+	-3	!
3 !									!
4 !									!
5 !	Equation 2:	3		y =	-1	x	+	7	!
6 !									!
7 !									!
8 !									!
9 !		Equation 1			Equation 2				!
10 !	x value	y value		x value	y value		Difference		!
11 !	=====================			====================			============		!
12 !	-2	-1.4		-2	3.000000		4.400000		!
13 !	-1	-1		-1	2.666667		3.666667		!
14 !	0	-.6		0	2.333333		2.933333		!
15 !	1	-.2		1	2		2.2		!
16 !	1.5	0		1.5	1.833333		1.833333		!
17 !	2.7	.48		2.7	1.433333		.9533333		!
18 !	3	.6		3	1.333333		.7333333		!
19 !	4	1		4	1		0		!

Fig. 6 Typical verbal problem

	A	B	C	D	E	F	G	H	I	J	
1 !	A	B	C	D	E	F	G	H	I	J	!
2 !											!
3 !	Problem: The owner of a clothing store bought 12 shirts and 6 hats for $210.										
4 !	A week later, at the same prices, she bought 9 shirts and 4 hats for $150.										
5 !	Find the price of a shirt and the price of a hat.										
6 !											
7 !	Let s = cost of a		12s	+	6h		=	120			
8 !	shirt in										
9 !	dollars										
10 !											
11 !	Let h = cost of a		9s	+	4h		=	150			
12 !	hat in $.										
13 !	---										
14 !	Guess for		Guess for		12s + 6h			9s + 4h			
15 !	Cost of		Cost of								
16 !	One Shirt		One Hat		($210)			($150)			
17 !	==========		==========		==========			==========			
18 !	5		10		120			85			
19 !	10		10		180			130			
20 !	15		15		270			195			
21 !	13		13		234			169			
22 !	11		11		198			143			
23 !	12		12		216			156			
24 !	11		13		210			151			
25 !	10		14		204			146			
26 !	10		15		210			150			
27 !	0		0		0			0			
28 !	0		0		0			0			
29 !	0		0		0			0			

Solving systems of equations

Figure 5 shows a template that leads students to the solution of a system of equations. By keeping the x values entered by students for equation (1) the same as those for equation (2), students solve the system through systematic trial and error by trying to minimize the difference between the y values in the two equations. Thus, the spreadsheet offers students a method of solution that increases their sense of the numerical relationships. After solving several problems using systematic trial and error, students often also recognize both the need and the rationale for a more direct approach to the solution. The teacher can use this same design to solve multiple systems just by changing the coefficients in the equation cells and can investigate a range of problems including inconsistent and dependent systems. A very similar approach can also be applied to the solution of a typical verbal problem as illustrated by figure 6.

Solving quadratics

Figure 7 shows a template that is designed to help students determine the roots of a given quadratic equation. Here students are given a table of values for the specified quadratic and have the ability to change both the starting value for the table and the increment. Thus, as in the example shown, students can see a change of sign between x = -3 and x = -2 and another between 3 and 4. Noting these changes,

students can then create new tables getting successively closer to the roots. Figure 8 shows the same template after several student changes of starting value and increment, yielding the value of x= -2.4 for the root between -3 and -2. Students would then repeat the process to locate the second root. Since the same template can be used for any quadratic just by changing the coefficients, teachers can have students investigate integral, irrational, or even imaginary roots by entering different coefficients in cells D3, F3, and H3. This design can also be used to find roots of cubic or higher-order equations and to create graphs with the appropriate spreadsheet.

Powers, roots, and logarithms

Using spreadsheets to develop number awareness is especially useful with such unfamiliar concepts as powers, roots, and logarithms (see fig. 9). Students often have no experience with raising numbers to nonintegral powers, but the spreadsheet can offer that experience, which might be critical to students' understanding of logarithms. Very similar templates set up so that students need only enter their conjectures to such questions as "What number when raised to the twelfth power will give you 0.3?" "What number when raised to the twenty-fifth power will give you 6 million?" or "What root of 1000 would be closest to 5?" allow students to experiment with numbers and their properties.

Fig. 7 Locating roots of quadratics

	A	B	C	D	E	F	G	H	I
1									
2									
3			Quadratic:	1	$x^2 +$	-1.1	$x+$	-8.4	$= 0$
4									
5	x	$f(x)$							
6	=====	======							
7	-5	22.1							
8	-4	12	Starting value $=$			-5			
9	-3	3.9							
10	-2	-2.2							
11	-1	-6.3							
12	0	-8.4	Increment value $=$			1			
13	1	-8.5							
14	2	-6.6							
15	3	-2.7							
16	4	3.2							
17	5	11.1							
18	6	2.1							

Trigonometry

Figure 10 shows a template designed to allow students to identify or verify a Pythagorean trigonometric identity. The figure shows what the computer screen might look like after a student had entered several angle measures in the first column. If a teacher wanted to make the pattern less obvious to students, the right-hand column could be deleted. Similar templates can be used to allow students to recognize or verify other trigonometric or algebraic identities. Since most spreadsheets also contain logical operators, a similar procedure can be used to verify logical (i.e., truth table) statements as well.

Additional Techniques and Applications

Although the examples cited require only that students enter numerical values, having students create either part or all of the spreadsheet also has advantages. To do so, students need to be sufficiently knowledgeable about the mathematics to be able to set up the appropriate algorithms and formulas. For example, to have students use the spreadsheet to create their own trigonometric tables requires that students know how to convert from degrees to radians and to use the trigonometric functions available.

Since many spreadsheets have sine, cosine, and tangent functions available, students who are asked to design a table of all six functions must understand their interrelationships. Students who are also asked to set up the table to print out values for which the user can designate starting value and the increment must use other fundamental algebraic relationships

of variables and increments. A significant additional result is that students seem to have pride in using a trigonometric or logarithmic table that they have made.

The spreadsheet can also be used as a powerful calculating tool to explore the effects of extreme values on the mean, compare the computation of standard deviations using different formulas, or compare approximations of the area under a curve by altering the number of intervals. The spreadsheet can be as powerful a problem-solving tool as a computer program or a calculator. Once familiar with spreadsheet commands, students can be asked to use the spreadsheet to calculate factorials, generate Pythagorean triples, or produce the Fibonacci sequence. At any level of involvement, however, an integration of the spreadsheet into the mathematics classroom will not only allow students to add to their repertoire of problem-solving techniques, but also give experience with a tool that can be applied.

Appendix: Formulas for the Figures

Formulas for figure 1
A9: +E3
A10: +A9 + E4
A11: +A10 + E4 (replicated through A29)
B9: +360/A9
B10: +360/A10 (replicated through B29)
C9: + 180 - B9
C10: + 180 - B10 (replicated through C29)
D9: + C9 * A9
D10: + C10 * A10 (replicated through D29)

Fig. 8 Finding the Root

```
 1 !  A  !   B   !     C      ! D !  E    ! F  !  G  ! H ! I  !
 2 !_____  _____  _____  ___  _____  ____  _____  ___  ___!
 3 !       ! Quadratic:  ! 1 ! x² +  ! -1.1! x+  !-8.4 ! = 0 !
 4 !                                                           !
 5 !   x      f(x)                                             !
 6 !=====  ======                                             !
 7 ! -2.5    .6                                               !
 8 ! -2.4    0      Starting value  =        -2.5             !
 9 ! -2.3   -.58                                              !
10 ! -2.2   -1.1                                              !
11 ! -2.1   -1.7                                              !
12 !  -2    -2.2    Increment       =         .1              !
13 ! -1.9   -2.7                                              !
14 ! -1.8   -3.2                                              !
15 ! -1.7   -3.6                                              !
16 ! -1.6   -4.1                                              !
17 ! -1.5   -4.5                                              !
18 ! -1.4   -4.9                                              !
```

Formulas for figure 2

A8-A25 blank or @NA until students enter values
B8: + 360/A8
B9: +360/A9 (replicated through B29)
C8: + 180 - B8
C9: + 180 - B9 (replicated through C29)
D8: + C8 * A8
D9: + C9 * A9 (replicated through D29)

Formulas for figure 3

A8: +(BD100 * B8 * B8 * B8) + (BE100 * B8 * B8) +
    (BF100 * B8) + BG

A9: +(BD100 * B9 * B9 * B9) + (BE100 * B9 * B9) +
    (BF100 * B9) + BG (replicated through A17)

Note: No formulas for figure 4

Formulas for figure 5

B12: +((E2/C2) * A12 + (G2/C2))
B13: +((E2/C2) * A13 + (G2/C2)) (Replicated through
    B23)
D12: +A12
D13: +A13 (replicated through D23)
F12: + ((E5/C5) * D12 + (G5/C5))
F13: +((E5/C5) * D13 + (G5/C5)) (replicated through
    F23)
H12: +F12 - B12
H13: +F13 - B13 (replicated through H23)

Formulas for figure 6

G18 + (12 * B18) + (6 * D18)
G19: + (12 * B19) +6 * D19 (replicated through
    G39)
I18: + (9 * B18) + (4 * D18)
I19: + (9 * B19) + (4 * D19) (replicated through
    I39)

Formulas for figures 7 and 8

A7: +F8
A8: +A7 + F12
A9: +A8 + F12 (replicated through A18)
B7: +D3 * (A7 * A7) + F3 * A7 + E3
B8: +D3 * (A8 * A8) + F3 * A8 + H3 (replicated
    through B18)

Formulas for figure 9

C4: +A4^B4
C5: +A5^B5 (replicated through C19)

Fig. 9 Exploring Exponents

	A	B	C	D
1				
2				
3	Base	Exponent	Value	
4	10	0	1	
5	10	1	10	
6	10	2	100	
7	10	3	1000	
8	10	-1	0.10	
9	10	-2	0.01	
10	10	.4771	3	
11	10	1.4771	30	
12	10	2.4771	300	
13	10	3.4771	3000	
14	10	-.5229	0.30	

Fig 10. Discovering trigonometric identities

	A	B	C	D	E	F
1						
2	x deg	sin x	cos x	sin² x	cos² x	sin² + cos² x
3	=====	=====	======	======	======	=============
4	30	0.500	0.866	0.250	0.750	1.00
5	42	0.669	0.743	0.448	0.552	1
6	26.5	0.446	0.895	0.199	0.801	1.00
7	145	0.574	-.82	0.329	0.671	1.00
8	212	-.53	-.85	0.281	0.719	1
9	415	0.819	0.574	0.671	0.329	1
10	-15	-.26	0.966	0.067	0.933	1
11	45	0.707	0.707	0.500	0.500	1
12	300	-.87	0.500	0.750	0.250	1.00
13	5246	-.44	-.90	0.192	0.808	1.00

Formulas for figure 10

A4: Initially @NA or blank (replicated through A13); then student entered values

B4: @sin(A3 * 3.1416/180) (replicated through C13)

C4: @cos(A4 * 3.1416/180) (replicated through D13)

D4: + B4 * B4 (replicated through E13)

E4: + C4 * C4 (replicated through F13)

F4: + D4 + C4 (replicated through G13)

Reference

O'Connor, Vince. *Spreadsheets: Applications for Algebra.* Handouts for a presentation at the 63d Annual Meeting of the National Council of Teachers of Mathematics, San Antonio, Texas, April 1986.

Bibliography

Arad, Ofar S. "The Spreadsheet: Solving Word Problems." *Computing Teacher* 14 (December-January 1986-87):13-15, 45.

Arganbright, Deane E. "The Electronic Spreadsheet and Mathematical Algorithms." *Two-Year College Mathematics Journal* 15 (March 1984):148-57.

————. "Mathematical Applications of an Electronic Spreadsheet." In *Computers in Mathematics Education,* 1984 Yearbook of the National Council of Teachers of Mathematics, 184-93. Reston, Va.: The Council, 1984.

————. *Mathematical Applications of Electronic Spreadsheets.* New York: McGraw-Hill, 1984.

Bratton, George. "Spreadsheets and Synthetic Division." *Computing Teacher* 15 *(March* 1988) :38- 40.

Brown, Joan Marie. "Spreadsheets in the Classroom." *Computing Teacher* 14 (December-January 1986-87) :8-12.

————. "Spreadsheets in the Classroom: Part II." *Computing Teacher* 14 (February 1987):9-12.

Hsiao, Frank S. T. "Micros in Mathematics Education—Uses of Spreadsheets in CAL." *International Journal of Mathematics Education, Science and Technology 16* (1985) :705-13.

Kelman, Peter, Art Bardige, Jonathan Choate, George Hanify, John Richards, Nancy Roberts, Mary Tornrose, and Joseph Walters. *Computers in Teaching Mathematics.* Reading, Mass.: Addison-Wesley Publishing Co., 1983.

Matheny, Art. "Simulation with Electronic Spreadsheets." *BYTE* 9 (March 1984) :411-14.

Minnesota Educational Computing Consortium. (MECC). *The Electronic Spreadsheet.* St. Paul, Minn.: MECC, 1983.

Peasey, Derek. "Using Spreadsheet Programs for Mathematics Education." *Micromath* 1 (Spring 1985) :44-46.

Russell, John C. "Probability Modeling with a Spreadsheet." *Computing Teacher* 15 (November 1987): 58-60.

Scott, Richard, and Roger Day. *A Tool for Applications and Algorithms: The Electronic Spreadsheet.* Handouts for a presentation at the 64th Annual Meeting of the National Council of Teachers of Mathematics, Washington, D.C., April 1986.

Spero, Samuel W. "Spreadsheets in the Classrooms." *Creative Computing* 11 (October 1985):88-90.

Tinker, Robert. "Spreadsheet Math...an Example of Educational Applications of General Software Tools." *Hands-On* 7 (Spring 1984):19-21.

Wood, John, Heather Scott, Steve Wright, N. Grandgenett, and Barry Newman. " Spreadsheets" *Micromath* 3 (Summer 1987) :36-50.

CHAPTER 3

Data Base Programs in the Mathematics Classroom

Ours is often called the "information society." Information that would never have been gathered or would have been lost 10 or 20 years ago is now saved and used for many purposes. Detailed census information, business statistics, medical records, tax records, opinion data, consumer trends, and so on are gathered and maintained by various groups. In the past, it would have been impossible to store, manipulate or make sense of all this information. But with computers, not only can we store huge amounts of information, easily retrieve it, and manipulate it, but we can use it for making very important decisions—whether to increase the prime lending rate, raise or lower the price of our product, support a certain piece of legislation, or change the instructional program of a pupil with learning problems.

The most important computer-based tool for storing and manipulating information is the *data base manager (DBM)*. This tool allows us to set up a format (template) for storing a particular set of data. For example, if we were setting up a data base of children's physical fitness information, the format of the data base might consist of the field names (a *field* is part of the data base that can contain a single piece of information) shown in Fig. 3.1.

```
Student:
Teacher:
Birthdate:
Weight 9/14:
Weight 1/15
Weight 5/29:
Height 9/14:
Height 1/15:
Height 5/29:
50 Yard Run:
880 Yard Run:
Situps:
Obstacle Course:
```

Fig. 3.1: Data Base template with 13 fields

Once the format of the data base has been established, the data itself can be entered. Data about each individual is entered into a *record*. Fig. 3.2 shows two records from a data-gathering/metric-measurement activity (see Pupil Activity 3.2. and Figures 3.9 and 3.10).

```
Last Name: Jones
First Name: Sylvia
Birthdate: 2/15/80
Sex: Girl
Eye Color: Blue
Hair Color: Brown
Favorite TV Show: Simpsons
Favorite Singer or Group: Michael Jackson
Favorite Kind of Animal: Cat
Favorite Sport: Soccer
Favorite Color: Magenta
Height (cm): 151
Hand Span (cm): 18
Length of Foot (cm):20
Area of Hand (sq. cm): 112
Arm Span (cm): 120
Weight of Shoe: 160
Weight of Pencil: 15
```

```
Last Name: Sanchez
First Name: Hector
Birthdate: 12/5/79
Sex: Boy
Eye Color: Brown
Hair Color: Brown
Favorite TV Show: Football
Favorite Singer or Group: Madonna
Favorite Kind of Animal: Frog
Favorite Sport: Football
Favorite Color: Blue
Height (cm): 147
Hand Span (cm): 20
Length of Foot (cm): 23
Area of Hand (sq. cm): 125
Arm Span (cm): 131
Weight of Shoe: 221
Weight of Pencil: 8
```

Fig. 3.2: Two records of student data

Before it is inputted, however, the format of the descriptors stored in the fields must be established. This is important, because if different codes or words are used for the same item, the selecting or sorting of that item will be inaccurate.. For example, if in the "Sex" field, you used the codes F, FEM, GIRL, FEMALE, and G to represent girls, it would be difficult to select the records of all the girls in the data base—especially if your data base was large—if you had forgotten some of the descriptors you used. When selecting records with a particular attribute, you first name the field from which you are selecting (e.g., "Sex"), then you select the type of comparison to be made (e.g., "Equals," "Greater Than," "Less Than," "Contains"), and then you name the descriptor (e.g., "Female"). See Fig. 3.3.

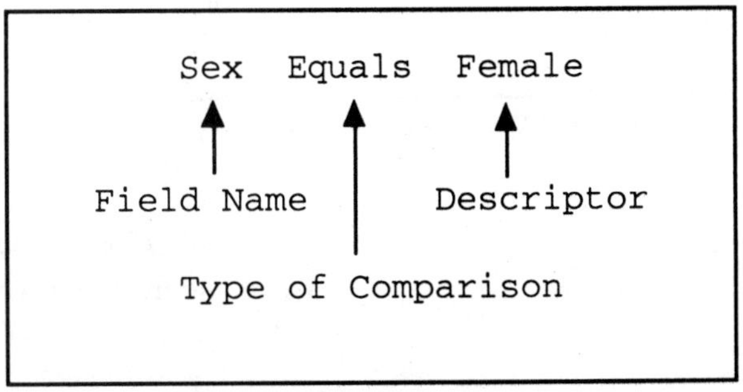

When you have described the type of selection you wish to make, the computer searches in the specified field (e.g., "Sex") throughout the entire database for the records which meet the criteria you have established. The records that meet the specified criteria are placed in a subset that you can examine, manipulate further, or format for printing.

Fig. 3.3: *Selecting a subset of records from a data base*

There are many possible types of selections you can make. You could select all records in which the field values are greater or less than a certain value (e.g., "Hand Span Less than 15"), or that contain specified characters (e.g., "Favorite TV Show Contains Cosby"). Most DBMs allow for many other types of selections as well, including examining and selecting from several fields at once. For example, if you wanted to select a group of children from the data base with blond hair and blue eyes who have big feet, you would select records as follows: "Hair Color Equals Blond *and* Eye Color Equals Blue *and* Length of Foot Is Greater Than 25."

Data base managers allow other types of searches and manipulations. You may ask the DBM to "Find" the record(s) containing a certain sequence of characters. For example, suppose that your student data base contained 175 records and you wanted to see the information about "Lisa," but you did not remember her last name. You could use the "Find" feature of the DBM to locate any record containing that name.

Data base managers allow the user to arrange the records in your data base in ascending or descending alphabetical or numerical order using the descriptors in any field. For example in the student data base, you could arrange the records in alphabetical order using the last name field. Or you could arrange them ascending order according to the students' height.

Most DBMs will allow you to display your data on the screen in two different formats—either table or record format. The format shown in Figure 3.2 is *record format*. When data is displayed in *table format*, it resembles a spreadsheet as in Fig. 3.4. Note, however, that data displayed in table format cannot be manipulated like a spreadsheet. Usually, only the first few fields of each record are displayed. Also, only the most powerful of DBMs will allow you to do computations more sophisticated than adding the values in a particular field of the data base.

Last Name	First Name	Sex	Birth Date	Teacher	Weight 9/14	Height 9/14
Adams	Julie	F	8/3/79	Rubins	59	52
Armosa	Philippe	M	12/7/80	Rubins	56	54
Azalone	Jacob	M	2/18/80	Cacciolfi	73	58
Chin	Julie	F	4/8/80	Smith	55	51
Chilson	Millie	F	1/5/80	Cacciolfi	62	55
Clayborn	Fred	M	3/23/80	Rubins	66	57

Fig. 3.4: Data base in table format.

Most DBMs allow you to design either table-style or record-style reports from the information in your data base. When you wish to design a report in the AppleWorks Data Base Manager, for example, you enter an editor that allows you to select the fields to include in the report. If you are preparing a table-style report, you can select the columns (fields) to display, their width, and various other attributes. If you are preparing a record-type report, you may select and position the desired fields on the page as desired. This allows you to do anything from printing labels (only using the name and address fields of a larger data base, for example, and specifying that your "page" length be only 6 lines) to reports containing complete information in the data base. Figure 3.5 is a record-style report based on a physical fitness data base (note that you may choose not to display the field names if you so desire).

```
Washington School, Fifth Grade Physical Fitness Report   6/5/90

Student:  Julie Adams              Birthdate: 8/3/79
Teacher:  Rubins
Weight 9/14:    59        Height 9/14:    56
Weight 1/15:    62        Height 1/15:    57.25
Weight 5/29:    67        Height 5/29:    61.5
50 Yard Run: 13.5
880 Yard Run: 437
Situps: 8
Obstacle Course: 29


Student:  Philippe Armosa
Teacher:  Rubins
Weight 9/14:    56        Height 9/14:    61
Weight 1/15:    68        Height 1/15:    63
Weight 5/29:    73        Height 5/29:    64
50 Yard Run: 10.3
880 Yard Run: 401
Situps: 23
Obstacle Course: 22
```

Fig. 3.5: Report with physical fitness data base.

Student Use of Data Base Managers

In the classroom, we have tended to give students exercises or problems in math that are neither long nor complex—operating on two numbers or solving problems that are stated in a couple of sentences and involve one or two operations. In real life, however, we must often solve problems that involve large amounts of data. Doing income taxes, recarpeting your house, and planning a trip are examples. Somehow, we have hoped that children will be able to make the transition from doing short, simple math tasks to the more complex problems faced in real life. That is not necessarily the case. The National Assessment of Educational Progress (Kouba, 1988) has repeatedly shown that many children are lost when trying to solve problems involving more than one operation. The National Council of Teachers of Mathematics, in their Curriculum and Evaluation Standards for School Mathematics (1989), has recommended that data gathering, organizing and analysis be an important part of the elementary mathematics curriculum.

Data base managers are well suited for data organization and analysis in the classroom. Data in mathematical and integrated (e.g., math/social studies) problem-solving activities are often textual as well as numeric—both of which are readily handled by DBMs. Many data bases managers such as AppleWorks, Microsoft Works, and pfs:file report are simple enough that they can be used for powerful activities by elementary students. Bank Street Filer is a data base manager that was designed for use in the classroom. In addition, a number of companies (e.g., Sunburst, Scholastic) produce premade data bases—primarily in the area of social studies—which are ideal for mathematical problem-solving activities.

Data base managers may be used at different levels in the classroom, ranging from the manipulation of a premade data base to creating a data base from scratch. Early experiences with data bases should involve the manipulation of data bases made by others for solving problems so that children learn what a data base is and how it may be manipulated. These experiences will be of great value if they are later asked to enter data into a template or to design their own data base—they will know what a DBM is and how it is used, and how data may be structured and described. The general sequence of introducing data bases into the classroom is as follows:

1. Students work with pre-made data bases—sorting, selecting, and so on, to solve problems.
2. Students gather, enter, and manipulate data using pre-designed data base template, then manipulate the data base to solve problems.
3. Students (with the help of the teacher) design data base and enter and manipulate data.

Note that, while it might be desirable for all children to eventually create a data base from scratch, the progress you make toward that goal depends upon the type of activities you do, the level of your students, and the amount of assistance you give them. Most fourth and fifth graders, for example, would need a great deal of assistance to design and create a data base (step 3), while they could be quite successful at gathering data and entering it into a pre-made template (step 2). Average and above students at the junior high and high school levels would have little difficulty progressing to step 3, especially when working in groups containing one or more computer-literate students.

Introducing Children to Data Bases: Using Pre-made Data Bases

A variety of publishers produce premade data bases for AppleWorks, PFS: File, Microsoft Works, and Bank Street School Filer. These can be used by students to solve mathematical problems.

Pupil Activity 3.1: Using a pre-made data base

Objective
Solve problems using a data base with numeric and textual information.
Grouping
Small groups or pairs of students.
Materials
A DBM (e.g., AppleWorks, Microsoft Works); A pre-made data base containing data about states, provinces, or countries; Computers.
Procedure
Obtain a pre-made data base containing information about the states in the United States or provinces in another country, or a data base containing information about countries (e.g., U.S. Database sold by Sunburst, World Geography sold by Scholastic). If you do not have access to such a data base, create one using information from the World Almanac or similar reference book. Set up the data base with the following fields: (1) State (name); (2) Population; (3) Urban population (percent); (4) Area (square miles); (5) Top industry; (6) Second industry; (7) Top crop; (8) Second crop; (9) Per capita income; (10) Unemployment (percent); and (11) Education (amount spent per pupil). *Note*: Fields 2, 3, 4, 9, 10, and 11 should be set up as numeric and not as text fields. Show the children how to sort the data base alphabetically and numerically, and how to select records. Depending on the background of the students, you may need to review or introduce the concept of percentage and the meaning of percentage numerals. Next, ask them to do the following activities (or similar activities that are relevant to the data base you are using):
A. Sort the states according to population. Which 5 states have the largest population?
B. In which states do less than 50 percent of the population live in cities?
C. In which states is wheat the top crop?
D. In which states is oil the main industry?
E. Which five states have the highest unemployment?
F. Which are the largest states?
G. Which states are more than 80% urban and spend less than $3000 a year per pupil on education?
H. In which states is agriculture either the first or second industry?
I. Which states have a population of more than 8 million and are less than 60% urban? Same question, but more than 80% urban?

These explorations could lead to discussions about the relationship between the size, population, or major industries in a state and unemployment or expenditures for education.

Another possibility is to gather data about your pupils, create a data base with the information, then have the students manipulate it. For example, the data might be obtained from a questionnaire involving information that is of interest to them and also involving measurement data.

Pupil Activity 3.2: Student Data Base

Objective
Measurement in centimeters, square centimeters, kilograms; Solve problems using data base manager.

Grouping
Individual, small group, large group.

Materials
Pencil and paper, questionnaire, metric measurement devices; DBM and computers.

Procedure (Instructions to Teacher)
Create a questionnaire about personal characteristics of your students including measurement activities (hand span in centimeters, circumference of ankle, length of pace, area of handprint in sq. cm, etc.), and including nonmeasurement data (eye color, hair color, favorite TV program, favorite singer, etc.). See Fig. 3.6 for an example. Administer the questionnaire to students in your class and have them do the appropriate measurement activities so that they can complete the questionnaire. Next, create an appropriate template and enter the data. After teaching students to use the DBM, have them manipulate the data base to answer questions about the data. See Fig. 3.7.

The article by Nancy Edwards, Gary Bitter, and Mary Hatfield at the end of this chapter describes data base activities using data from the log of a ship bringing immigrants to the United

```
Directions: Provide all the information requested below.
The information you provide will be used in a computer
activity in a few days.
Name_____
Birthdate_____
Sex_____
Eye Color_____
Hair Color_____
Favorite TV Show_____
Favorite Singer or group_____
Favorite Kind of Animal_____
Favorite Sport_____
Favorite Color_____
Height (cm) _____
Hand Span (cm) _____
Foot Length (cm) _____
Area of Hand* (sq. cm) _____
Arm Span (cm) _____
Weight of your shoe (gm) _____
Weight of your pencil (gm) _____
```

*The easiest way to find the area of your hand is to place it on a sheet of centimeter grid paper (fingers together), trace the outline of your hand, then count the number of square centimeters inside the region.

Fig. 3.6: Questionnaire for data base activity

```
Directions: Use AppleWorks and the data base called
Student Info to answer the questions below.

Which students have blue eyes?

Which students have black hair?

Whose favorite TV show is The Cosby Show?

How many people's favorite animal is cats?
Who is taller than 130 cm.

What is the total of all the hand spans in the class?
What is the average foot length in this class?
Whose feet are longer than average in this class?

Who has the heaviest shoe in this class?
If there are 1000 grams in a kilogram, what is the weight
in kilograms of all the shoes in this class?
Write the names of the 5 people with the heaviest pencils
in this class.
```

Fig. 3.7: Data base activity questions

States during the 1800s. Students could solve the problems described in the article after the teacher creates the template and enters the data shown in the article.

Entering and Manipulating Data with Premade Template

The next step is to have children gather and enter data where the template has been predesigned and set up by the teacher. Of course, before students enter the data, you will need to thoroughly discuss with them the need for consistent descriptors. For a particular data-gathering/analysis activity, you may wish to decide as a class what the descriptors will be, then write them on the chalkboard for all to see (e.g., the descriptor for "full sunlight" will be SUN). Note that since most children are not accomplished typists, it is best to keep the descriptors short.

Pupil Activity 3.3: Bean Sprouts

Objective
Gather data and evaluate data from science experiment.
Grouping
Whole class.
Materials
27 styrofoam cups, planting soil, bean seeds; plant growth solution, water; locations with different lighting conditions; DBM with template set up for evaluating the results of the experiment—the following fields should be in the template: Lighting, Water per day, Fertil. per week, 2 week height, 4 week height, 6 week height, Plant color (6 weeks) Plant appearance (6 weeks).
Procedure
Investigate the effects of lighting, water, and fertilizer on the growth of bean plants. Plant 27 bean seeds, each in a styrofoam cup containing planting soil. Place groups of 9 cups in three different lighting conditions (e.g., in a cupboard, back of the classroom away from windows, next to the windows). In these groups, 3 each will receive 1 ml of water a day, 3 will receive 5 ml a day, and 3 will receive 10 ml a day. In each group of 3, 1 will receive 0 drops a week of plant food, 1 will receive 3 drops a week and one will receive 6 drops a week. Note that there is 1 bean (in a styrofoam cup) for each lighting, water and fertilizer condition. Enter the data about the different plants into the data base (it will have already been set up by your teacher). Discuss the effect of water, light, and fertilizer on the growth of bean plants. Which combination of water, light, and fertilizer produces the best plants?

Pupil Activity 3.4: Polygons Data Base
Objective
Observe the properties of polygons; store and manipulate polygon data using DBM.
Grouping
Pairs or small groups.
Material
DBM

Procedure

Set up a data base template with the following fields: Name of figure, Number of sides, Description, Area formula, Perimeter formula, Total interior degrees. Introduce the activity by defining *polygon*, and entering data for two or three different kinds of polygon into the data base (e.g., triangle, square). Students then search for different polygons (perhaps using mathematics books and other reference materials)—such as trapezoids, rectangles, parallelograms, quadrilaterals, and regular pentagons—in small groups and enter information about them into their data base as they find them. If you wish, set up a master data base in which you enter information about each new polygon that is found and the name of the person or group finding it. After the data base contains numerous polygons, have the students manipulate it to answer questions: Which polygons have four sides? Which figures have interior angles totalling less than 500 degrees? See the article by Marilyn Ford in this chapter for a description of a similar but much more elaborate activity.

Creating a Data Base from Scratch

The final step in learning to use data bases is to have students design their own. A good choice for the first experience in designing a data base is Attribute Blocks or Animal Attribute Tiles[3], not only because they have a limited number of attributes (fields) but because they can be used in mathematical logic activities. You might wish to start by discussing how the Attribute Blocks are different from one another, then, as a group, determine the different attributes of the set of pieces. Attribute Blocks may be purchased in several different configurations, depending on the manufacturer. A typical set will have 32 blocks with three or four attributes: color (e.g, blue, yellow, green), size (e.g., small and large), shape (e.g, square, circle, diamond), and/or thickness (e.g., thick, thin) as attributes. Do a variety of activities with the Attribute Blocks (e.g,. have children find all blocks that are squares; have them find all blocks that are small; have them find all blocks that are circles and yellow; have them find all shapes that are thin and green or yellow). Next, as a group, design the data base. Decide what the fields in the data base will be, enter the "Data base design" (or "Category create") mode and create the needed fields—one field per attribute.

Next, make copies of the "template" for pairs (or small groups) of students. Give each group a set of attributes and have them make one record for each attribute piece, using the appropriate descriptors. Fig. 3.8 contains one record from the Attribute Block data base. Thus, if you have 32-piece Attribute Block sets, your data base will have 32 records.

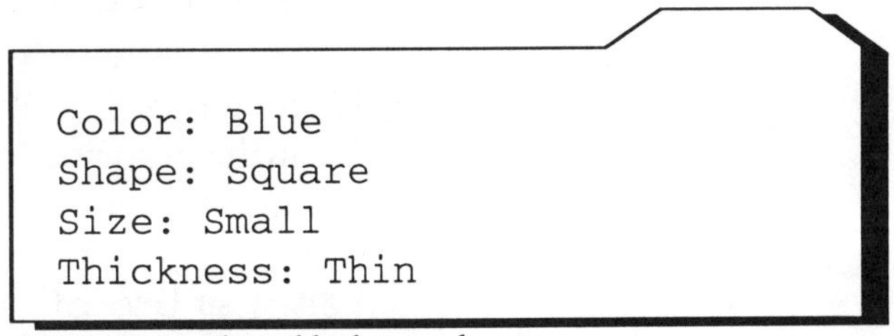

```
Color: Blue
Shape: Square
Size: Small
Thickness: Thin
```

Fig. 3.8: Attribute block record

---

[3] Attribute Blocks and accompanying student workbooks are available through local teacher-supply stores or through mail-order companies such as Dale Seymour Publications, P.O. Box 10888, Palo Alto, CA 94303 or Creative Publications, 5040 West 111th Street, Oak Lawn, IL 60453. You could also make them yourself from tagboard or colored plastic sheets.

After data about the Attribute Blocks have been entered into the data base, have students manipulate it to answer questions about the Attribute Blocks. You may wish to have students manipulate the data base and use the blocks themselves to verify their findings. Note that if students are using the data base to find shapes that are green, they would enter "Color equals green" when using the data base to make selections, or they would enter "size equals small and shape equals triangle" to find all shapes that are triangles and small. Here are some typical tasks:

1. Find and describe all green blocks.
2. Find and describe all triangles.
3. Find and describe all shapes that are small and thin.
4. Find and describe all shapes that are thick, large, *and* blue.

The following is a sample activity where children would create their own data base to investigate mathematical ideas or solve problems involving mathematics.

Pupil Activity 3.5: Christmas Shopping Spree

Objective
Gather data from catalog, use DBM.
Grouping
Pairs or small groups of students.
Materials
DBM, Sears Catalog or equivalent.
Procedure
Imagine that you live on a remote island and that you must do all your Christmas shopping using mail-order catalogs. You have saved $300 from doing odd jobs for your Christmas shopping. Set up a data base to help you plan your purchases. You should plan to purchase at least 10 gifts. Your data base should have the following categories: Gift name, Give To, Name of catalog, Order number, Price, Tax, Total. Print out the data base in table form, including the total cost of all the gifts, including tax.

The article at the end of this chapter by Marilyn S. Ford (1989), gives another example of how children might create a data base with geometric shapes. The activity involves helping children observe and develop precise language to describe plane geometric shapes, categorizing the shapes, recording information about them, and, finally, building and manipulating a computerized data base with all the information.

Teacher Use of Data Base Managers

A major part of a teacher's job is managing information. We maintain a large quantity of information about students: names, addresses, telephone numbers, test scores, and a variety of written comments. We may need to keep detailed records on the instructional programs for students with special needs. Or we have materials in our classrooms that need to be cataloged. We may be in charge of special events such as a school fund-raising project, a field trip, class play, or athletic event. If, in these types of situations, we have a set of information about a group of individuals, consisting of words and/ or numbers, where the same type of information is recorded for each individual, a computerized data base manager will greatly facilitate the task of organizing and manipulating the data.

Examples of the use of data base managers by teachers to keep track of information about their pupils were given in Chapter 1 and earlier in this chapter. A data base manager could also be used to keep track of materials in your classroom or school such as materials in a math lab (Fig. 3.9), computer software, textbooks, audiovisual materials, or individualized math programs.

Another way of organizing such a math materials data base would be according to the mathematical concepts and skills with which they would be used. They could then be selected as needed according to the desired attributes of the activity. For example, if you had set up the data base shown in Fig. 3.10 and wanted a whole-class activity involving addition of fractions, you would select records using the following selection criteria: "Grouping equals whole class and Concept/Operation contains addition and Concept/Operation contains fraction."

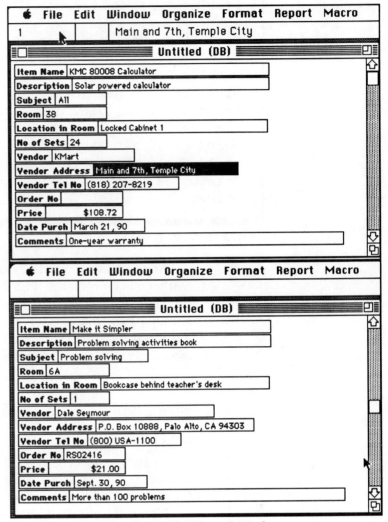

Fig. 3.9: Math lab data base with Microsoft Works

```
Name: Decimal Bingo
Type: Game
Grouping: Whole class
Duration: 45
Concept/Operation: Decimal Numerals
Materials: Bingo cards, chips, markers
Where stored: Math Lab, Shelf 1
Source: Self made
```

Fig. 3.10: Math Activity Data Base

```
Problem Type: Routine
Number Type:  Fraction, Unlike Den.
Operation Type: Addition, Subtraction
Problem Features: Extra Info., Multiple Oper.
Problem Source: HBJ Math 6, p. 145
Problem: Mrs. Wask baked an apple and a pumpkin pie and
invited Jim, Silvie, and Jason to help her eat some of it.
Jim ate 1/6 of the apple pie, Silvie ate 1/4 of the
pumpkin pie and Jason ate 1/3 of the apple pie.  What
fraction of the apple pie was left?
Strategies: Restate, Num. Sen.
Answer: 1/2
```

Fig. 3.11: Record from PS data base.

Another use of a data base manager would be to create data bases of mathematics problems (Fig. 3.11). Once created, you could easily select problems with specific attributes (e.g., Problem Type equals routine and Number Type contains whole number and Operation Type contains multiplication) and print them out for use by students. And, the data base can be modified and expanded as desired.

Bibliography

Bright, G. W. 1989. Data bases in the teaching of elementary school mathematics, *Arithmetic Teacher* 37(1, September): 38-42.

Curriculum and Evaluation Standards for School Mathematics. 1989. Reston, Va.: National Council of Teachers of Mathematics.

Edwards, N. T., Bitter, G. G., and Hatfield, M. M. 1990. Teaching mathematics with technology: Data base and spreadsheet templates with public domain software. *Arithmetic Teacher* 37(8, April): 52-55.

Ford, M. S. 1989. Databasing geometry in the elementary classroom. The Computing Teacher 17(1, August/September): 20-25.

Kouba, V. L et al. 1988. Results of the Fourth NAEP Assessment of Mathematics: Number, operations, and word problems. *Arithmetic Teacher* 35(8, April): 14-19.

Wiebe, J. H. 1988. *Teaching Elementary Mathematics in a Technological Age.* Scottsdale, Ariz.: Gorsuch, Scarisbrick.

Wiebe, J. H. 1990. Teaching mathematics with technology: Data base programs in the mathematics classroom. *Arithmetic Teacher* 37(5, January): 38-40.

Discussion Questions

1. With what kind of data would a spreadsheet be a more appropriate tool than a data base manager?

2. Why would a spreadsheet program be a more appropriate tool for managing a teacher's grades than a data base manager? Why would a data base manager be more appropriate than a spreadsheet for storing and manipulating information about pupils in a school or at a grade level (name, address, CTBS math score, etc.)?

3. Find a product review or product description of a powerful data base manager such as dBASE IV or Paradox. What capabilities does this package have which were not discussed in this chapter? Would any of these capabilities be of use to you as a teacher or to your students?

4. Would a data base manager be a useful tool in solving traditional word problems found in most elementary math textbooks? Why or why not?

5. Suggest some mathematics or integrated (science, social studies, etc.) mathematics data-gathering activities where a data base manager would be useful in organizing and manipulating the data.

Activities

Activity 3.1

Objective
Use a data base to manage and explore mathematical principles.
Rationale
This activity will require you to learn to set up and use a data base manager. It will also require you to recall and evaluate mathematical principles.
Materials
Data base manager, computer.
Advance Preparation
Read this chapter. You should know how to set up the categories (fields) in a data base, how to enter data into the data base, and how to prepare a report from the data base.
Procedure
Set up a data base of mathematical principles. Fig. 3.12 shows a sample record with appropriate fields. Do the following:

```
Principle: Commutative Property for Addition
 Operations: Addition
 Numbers: Real (including rational, integers, whole)
 Math Formula: A + B = B + A
 Example:  5 + 3 = 3 + 5  (the sum is 8)
 Description: The order in which you add numbers is irrelevant
 Uses: Order of addends can be changed when using calculator,
       students do not need to learn A + B and B + A as
       separate facts.
```

Fig. 3.12: Mathematical principles data base

1. Create the categories (fields) in the data base as shown in Fig. 3.12.
2. Find and enter at least ten mathematical principles into the data base.
3. Make a printout of all the records in the data base.
4. Arrange the records alphabetically, according to the principle.
5. Select those principles involving addition. Repeat for multiplication.
6. Select all principles involving multiplication or division.

Activity 3.2

Objective
Create a data base of problems; gain further experience in using a data base manager.

Rationale
Such a data base will be useful since all problems you might want to use in your classroom will be stored in one location and will be easily accessible, and you will be able to select and print the problems based on the desired criteria.

Materials
Data base manager; resource materials containing mathematics problems; computer.

Advance Preparation
Read this chapter. You should know how to set up the categories (fields) in a data base, how to enter data into the data base and how to prepare a report from the data base.

Procedure
Set up a problem-solving data base with categories (fields) similar to those shown in Figure 3.11. Using elementary math textbooks or other resource materials, collect 10 problems appropriate for students at the grade level you teach, or intend to teach. Enter them into the data base. Select and sort in several different ways (e.g., select all those records where Number Type contains fraction and Operation Type contains addition and Problem Features does not contain extra information) and print out the resulting records in table format.

Databasing Geometry in the Elementary Classroom

by Marilyn Sue Ford

Teachers have been quick to think of uses of databases to teach social studies and science. Within these contexts databases have also been used to explore mathematical situations. Are there areas in mathematics that can benefit from the organization that a database brings to a set of data?

The attributes of geometric figures might benefit from such organization if the students themselves develop the database and then use it. According to Burger (1988), elementary geometry instruction should reflect the research of van Hiele. Elementary students must be provided with opportunities to identify, analyze, and classify shapes through concrete to abstract reasoning experiences and eventually apply geometric properties to problem situations. An information processing computer environment facilitates and supports Burger's recommendations for a progressive, sequential, and comprehensive geometry curriculum in the elementary grades.

In this teaching unit for intermediate grades, students determine the geometric attributes of a variety of triangles and quadrilaterals and set up a database for these figures and their attributes. Then they use the database to answer geometric questions about the figures. Pictures of the shapes used on display in the room help students connect the shape with its abstract attributes listed in the database.

Figure 1 Geometric Characteristics

Rectangle:

4 angles	all right angles
all angles = 90 degrees	4 vertices
all angles equal	sum of angles = 360 degrees
4 sides	opposite sides parallel
2 pairs of equal sides	quadrilateral
4-sided polygon	etc.

Preparation

This geometry database unit incorporates both pre-computer (offline) and computer (online) activities that alternate between whole-group and small-group instruction. The computer activities in the teaching unit are developed with the Apple ProDOS version of Scholastic *PFS: File* software, but can be adapted for use with any database management software.

Activities are also designed for a self-contained classroom with one computer and approximately 30 students. For whole-group activities, a large demonstration monitor (or computer interfaced with a projection system) and printer are preferred, but optional.

Small-group computer activities necessitate use of a computer station, learning-center approach. You will need to create a computer station schedule whereby each small group has approximately 20-30 minutes of online time for computer activities.

Materials Needed: Ample supply of butcher paper, thick felt tip marking pens, and a package of 8" x 10" lined index cards.

Preparation: For each student, make several paper copies of a rectangle. For each group of three students, make several paper copies of one plane geometry shape. The following geometric shapes will adequately meet the needs for a class of 30 students: isosceles triangle, equilateral triangle, scalene triangle, acute triangle, right triangle, obtuse triangle, square, parallelogram, rhombus, and trapezoid.

Organizing Geometric Observations

The first objective is to determine the characteristics of the plane geometry shapes and then organize these observations. To begin, provide each student with a copy of the "class" shape (i.e., the rectangle). If students are familiar with the basic geometric components of a rectangle, they should readily recognize such characteristics as perpendicular lines, parallel lines, right angles. and opposite equal sides. If necessary, protractors and rulers can be used for verification of side and angle measurements. Encourage students to determine the geometric terms used to describe a rectangle and list them on a sheet of paper. Then make a class list of their observations on the chalkboard (Figure 1).

Provide each small group with several copies of one plane geometry shape that will become their group shape for the duration of the teaching unit. Have each small group select a recorder and repeat the above process with their group shape. Each recorder should label a sheet of butcher paper with the name of their group shape and list the geometric terms the group decides describe the shape. Post the butcher paper lists for easy viewing. Have students organize their observations by looking for similarities among the lists. For example, ask which terms describe line segments. On a large sheet of butcher paper labeled Geometry Word Bank Chart, sort these terms together as DESCRIPTORS and give them a CATEGORY label such as "sides." Continue this process, asking which terms describe angles, general names, specific names, and so forth.

Figure 2 Geometry Word Bank Chart

CATEGORIES	DESCRIPTORS
Types of sides	equal, nonequal, parallel, opposite, intersecting, perpendicular
Number of sides	# (numeric value)
Types of angles	obtuse, acute, right, equal
Number of angles	#
Sum of interior angles	#
Polygon name	triangle, quadrilateral
Shape name	isosceles triangle, equilateral triangle, scalene triangle, right triangle, acute triangle, obtuse triangle, rectangle, square, parallelogram, rhombus, trapezoid

If necessary, further refine categories. Either break them down into two or more categories or relabel them with more precise geometric terminology. The category "sides" could be broken down into two categories called "types of sides" and "number of sides". A category called "general names" could be relabeled as "polygon name." This process will produce a butcher paper Geometry Word Bank Chart similar to the one in Figure 2.

Extending Students' Geometric Knowledge Base

The descriptive, organizational process provides the classroom teacher with an informal assessment of students' geometric knowledge. Any areas not addressed on the Geometry Word Bank Chart become possible topics for instruction. For example, given that students can readily generate a list of

geometric terms such as those listed in Figures 1 and 2, then the next objective is to extend their geometric knowledge by exploring the geometric qualities of line and rotational symmetry (Renshaw, 1986) and tessellation. The ultimate outcome is to apply this newly acquired knowledge to each group shape and organize the information on the Geometry Word Bank Chart.

Provide each student with a copy of the "class" shape. Have each student cut out the rectangle. Define "line of symmetry" as an imaginary line which produces mirror images. Using a pencil, have students draw a possible line of symmetry on the rectangle. Fold the rectangle on the marked line. If the halves match, then the shape has line symmetry. Continue this procedure until all possible lines of symmetry have been found. Count each line/fold that produces matching halves as a line of symmetry.

Have the small groups refer to their plane geometry shape and apply the line of symmetry procedures to their shape. Groups should determine if their shape has none, one, or more lines of symmetry. Then have the whole group discuss how line of symmetry information could be added to the Geometry Word Bank Chart. Student discussion should be guided toward a category label such as "lines of symmetry" and descriptor such as "#" which indicates any numeric value equal to or greater than zero.

The preceding sequence of whole-group activity with the rectangle, small-group activity with group shapes, and whole-group discussion related to the Geometry Word Bank Chart should be repeated for exploring, applying, and organizing information on the rotational symmetry and tessellation qualities of geometric shapes. For rotational symmetry, cut out the shape, mark an "X" at the top, and a dot at the center of rotation. Place the cut-out shape on top of a tracing of the shape. Using a pencil point, hold the cut-out figure on the center of rotation. Rotate the cut-out shape about the center until it matches (fits) the shape on the bottom exactly. Keep rotating until the cut-out shape returns to the starting "X" position.

Figure 3 Amended Geometry Word Bank Chart

CATEGORIES	DESCRIPTORS
(data previously entered)	
Lines of Symmetry	#
Turns of Rotational Symmetry	#
Tessellates	yes, no

Count the number of rotational turns which create an exact match. The last turn, which places the cutout shape in the starting position, completes the cycle and does not count.

Cut out several copies of a geometric shape in order to determine the quality of tessellation. Arrange the shapes on a piece of colored construction paper according to the tessellation rule of touching edges. If the edges touch and completely cover the construction paper surface, then the shape tessellates a plane. If the edges touch, but leave spaces so that parts of the construction paper surface are still visible, then the shape does not tessellate a plane.

If further whole-group exploration of line and rotational symmetry and tessellation is warranted before small group activities, use a pentagon shape. The

Figure 4 Record Format for Types of Sides and Types of Angles

FIELD	DATA
Types of sides	all or # pairs of equal sides # pairs of opposite parallel sides # pairs of intersecting perpendicular sides
Types of angles	all or # pairs of equal angles # right angles # acute angles # obtuse angles

pentagon has five lines of symmetry, four turns of rotational symmetry, and does not tessellate a plane surface. These results will provide data which students can compare and discuss in relation to the rectangular outcomes. Once the exploration of symmetry any tessellation is completed for both the "class" shape and the group shapes, the resulting information must be added to the Geometry Word Bank Chart. Students' experiences and group discussions should produce an amended Word Bank Chart with the additions listed in Figure 3.

Creating a Database Record Format

The next step is to design a database record format using the Geometry Word Bank Chart as a guide. Post a large sheet of blank butcher paper next to the Geometry Word Bank Chart. Label the butcher paper "Plane Geometry Shapes File" with a subheading of "Record Format." Substituting database terminology, rename categories as *fields* and descriptors as

data. As a whole group activity, begin to construct a record format. Have the students discuss and list the fields in a logical sequence. For example, begin with name fields first, then side and angle, and end with fields related to other geometric qualities.

The exact phrasing of data is important, as all records must utilize the same data phrases and spellings for efficient and effective use of the database. Due to the nature of the topic, the two fields labeled "types of sides" and "types of angles" will probably be the most cumbersome to deal with. The key is to use the least number of words when selecting descriptive data. Reference to the "types of sides" category on the Geometry Word Bank Chart reveals the words equal, nonequal, parallel, opposite, intersecting, and perpendicular. Student discussion could lead to the following conclusions:

• describing some sides as equal means all other sides must be nonequal: therefore, use only *equal*.
• since the definition of perpendicular includes the word *intersecting*, use only *perpendicular*, and
• because the word *opposite* is used to define parallel sides, use only *parallel*.

Finalized Record Format

Using only the data words *equal*, *parallel*, and perpendicular, students must determine how to phrase data entry so all shapes can be accounted for. Using each group's shape, students should come to realize their shapes have either all, none, or pairs of equal sides; none or opposite pairs of parallel sides; and/or none or intersecting pairs of perpendicular sides. Thus the data section for "types of sides" will need to incorporate these ideas. The same process will need to be followed for the field "types of angles." Refer to Figure 4 for an example of how this process could result in a standard data format that all groups apply to their shape.

The whole group should now carefully review the format of all the fields and data. Some fields may have too many different kinds of data descriptors. If so, then these data descriptors could become fields of their own. Pursue further refinement into more fields with less data descriptors. This refining process will produce a butcher paper record format such as the one shown in Figure 5.

Using the Record Format chart, have the whole group develop a record for the "class" shape on another large piece of butcher paper. Stress preciseness, because the data descriptors must match those agreed upon for the Record Format chart. This large-group activity should result in a record for the rectangle which looks like Figure 6.

Figure 5 Plane Geometry Shapes File Record Format

FIELDS	DATA
Polygon name:	triangle, quadrilateral
Shape name:	isosceles triangle, equilateral triangle, scalene triangle, right triangle, acute triangle, obtuse triangle, square, parallelogram, rectangle, rhombus, trapezoid,
Number of sides:	#
Equal sides:	all, # pair(s)
Pairs of opposite parallel sides:	#
Pairs of intersecting perpendicular sides:	#
Number of angles:	#
Equal angles:	all, # pair(s)
Right angles:	#
Acute angles:	#
Obtuse angles:	#
Total interior degrees:	#
Lines of symmetry:	#
Turns of rotational symmetry:	#
Tessellates:	yes, no

Provide 8" x 10" index cards for each small group and have them copy the fields from the Record Format chart onto the card. Have each group refer to its plane geometry shape and complete an index card record for its group shape. Remember to stress that data descriptors must match those designed by the class. It is helpful to post both the Record Format chart and the Sample Record for the rectangle within easy viewing for all groups to use as a reference.

Creating the Computer Database File

The teacher or a knowledgeable student must now use the database software to create a file that incorporates the record format designed by the class. The file could be named "Plane Geometry Shapes." For ease of use, the blank computer record may be laid out differently from the butcher paper Record Format in order to use only one computer screen.

Adding Data to the Geometry File

This data entry activity will require one computer booted with the "Plane Geometry File" data disk. With the whole group demonstrate the data entry process using the "class" shape. In order to standardize the data format, refer to the prominently posted Record Format chart and the Sample Record chart for the rectangle as data is entered into each field. Once the record for the "class" shape is completed, carefully review each field for standard data format and potential spelling errors. If any data is incorrectly entered, retype the correct data before finalizing the adding process.

Post a computer station schedule which provides each group approximately 20-30 minutes of online computer time for data entry. Each small group will follow the same procedures outlined with the "class" shape to add its shape's data to the file. Groups should refer to their index card records while entering data into the computer database. Post the Record Format chart and the Sample Record chart for the rectangle near the computer station while students are entering their data to the file. Keeping the data format the same for all shapes will make it easier to compare and contrast records later on.

Simulating Database Searches

In order to explore a computer database students must understand two concepts. First, what is a computer database search? Second, how does a particular database accept search criteria? Understanding both ideas can be accomplished by having students simulate how a computer searches a database. Provide each small group with a hard copy printout of their computerized record from the "Plane Geometry Shapes" file. If a printer is not available, have groups use their index card record. Ask the whole group a geometric shapes question. For example, "Which geometric shapes have two equal angles?" Have each small group look at the hard copy or index card record that correlates with its group shape. If the shape answers the question, then someone in the group raises a hand. Have students tell which field they were looking at and what data they were looking for. (The field is "equal angles," and data is "2.")

Figure 6 Record for the Rectangle

FIELDS	DATA
Polygon name:	quadrilateral
Shape name:	rectangle
Number of sides:	4
Equal sides:	2 pair(s)
Pairs of opposite parallel sides:	2
Pairs of intersecting perpendicular sides:	4
Number of angles:	4
Equal angles:	all
Right angles:	4
Acute angles:	0
Obtuse angles:	0
Total interior degrees:	360
Lines of symmetry:	2
Turns of rotational symmetry:	1
Tessellates:	yes

Sample Record

Demonstrate how the computer database answers the same question. Computerized search criteria must match the data format used to create the file. Refer to the Record Format chart where data is specified as numeric by the number sign, or otherwise as alphanumeric. As the computer indicates a shape's record, have someone from that group stand up. Compare the raise of hands performed earlier to the students standing. The two groups should match. If any discrepancies occur, use the computer records and the hard copy printouts to carefully review the field(s) searched for any data format or spelling

errors. Update (change) incorrect records immediately and repeat the exercise. See Figure 7 for some sample geometric questions and search criteria strategies using the *PFS: File* database.

Search-criteria Strategies

Once students understand how to explore a database file, they are ready to apply search criteria strategies to solve other geometric shapes questions. As a whole group activity, formally introduce a step by step process for applying search-criteria strategies to answer questions about the geometric shapes. First, determine which fields will help answer the question. These fields should be written down. Then analyze which search criteria will be needed to search the field (i.e., equals, does not equal, is less than, etc.). Remember, search criteria must use the same data format as the Record Format chart. This search criteria should be written beside the appropriate field. The need to write down fields and search criteria becomes evident when a search goes astray. Being able to see what the computer was asked to do helps both you and your students analyze any searches that do not provide a reasonable answer. Finally, students should write down the information that appears on the screen in answer to the question. Other questions which require students to think about their answer and/or carefully scrutinize individual records can also be incorporated (see Figure 8).

Follow-up small group, online computer activities should stress use of this process for applying search criteria strategies. The computer station schedule established for data entry purposes can be followed for small-group exploration of the database search questions. After groups have answered the teacher-developed questions, encourage them to write and solve their own geometry question.

As a culminating activity, provide students with an opportunity to share their database shape questions that were written during the database search activity. Place questions on index cards at the computer station for groups to answer during online computer time.

Conclusion

According to Olds and Dickenson (1985) there are three stages for learning to use database management software in educational settings. The easiest is to use a database created by someone else while learning to search and print data. The next stage is to create and use a database given the record format with fields, but no data. As students collect and enter data they learn the importance of standardizing

Figure 7 Geometric Questions and Search Criteria

Exact match (Alphanumeric data)
Which geometric shapes have only one pair of equal sides?
TAB to the EQUAL SIDES field. Type: 1 pair(s)

Partial Match (Alphanumeric data)
Which geometric shapes belong to the quadrilateral family of polygons?
TAB to the POLYGON NAME field. Type: quad..

Numeric matches (Numeric data)
Which geometric shapes have no turns of rotational symmetry?
TAB to the TURNS OF SYMMETRY field. Type: 0

Which geometric shapes have two or more lines of symmetry?
TAB to the LINES OF SYMMETRY field. Type: >1

Which geometric shapes have fewer than four right angles?
TAB to the RIGHT ANGLES field. Type: <4

Numeric Range (Numeric Data)
Which geometric shapes have one to three acute angles?
TAB to the ACUTE ANGLES field. Type: = 1..3

Not matches (Alphanumeric or numeric data)
Which geometric shapes do not have "angle" anywhere in their names?
TAB to the SHAPE NAME field. Type: /..angle..

Which geometric shapes do not have interior angles which sum to 180 degrees?
TAB to TOTAL INTERIOR DEGREES field. Type: /180

Figure 8 The Search Process

Shapes Question
Which shapes have pairs of parallel and perpendicular sides? (Why are none of the shapes triangles?)

Fields and Search Criteria

PAIRS OF OPPOSITE PARALLEL SIDES: >0
PAIRS OF INTERSECTING PERPENDICULAR SIDES: >0

Answer: square, rectangle

(Although the right triangle has one pair of perpendicular sides, triangles do not have parallel sides.)

vocabulary. The most challenging stage demands that students create and use an original database from scratch. This geometry unit outlines a guided discovery process for developing the third and most difficult stage of databasing.

Moreover, this process incorporates a standard curriculum mathematical topic of geometry. While students identify, analyze and classify geometric shapes, they also learn about database management as a powerful computer tool for exploring and managing information. Computer literacy knowledge for today's students must include skills with information processing software. Coupling geometry instruction with a computer environment optimizes instructional time on task through the integration of mathematics and computer education.

References

Burger, W. F. (1988, November). One point of view: An active approach to geometry. *Arithmetic Teacher:* 36, 2.

Ford, M. S., & Gregg. L. (1987, March). The rockhound files. *Teaching and Computers:* 4, 10-15.

Olds, H. F. Jr., & Dickenson, A. (1985, October). Move over, word processors—here come the databases. *Classroom Computer Learning*: 6, 46-49.

Renshaw, B. S. (1986). Symmetry the trademark way. *Arithmetic Teacher*: 34, 6-12.

Woldman, E.. & Kalowski, P. (1985, February). Make your own dinosaur data base. *Teaching and Computers*: 2, 14-19.

Teaching Mathematics with Technology:
Data Base and Spreadsheet Templates with Public Domain Software
by Nancy T. Edwards, Gary G. Bitter and Mary Hatfield

Good programs are becoming available through public domain software, software that can be obtained at little or no cost whose duplication is legal. An example is the AppleWorks simulation program IMMIGRANT. It contains several programs, including the passenger lists of two ships from which the student is asked to "adopt" a family. The students help the family adjust to the housing, market, jobs and budgets of the 1840s and 1850s. IMMIGRANT yields numerical information in two formats: data base and spreadsheet.

Thought Strategies for Using Data Base Information

Data base information is a listing of pertinent details by categories. A student can use the data to gain an insight into the conditions, customs, physical needs, and types of people coming to the United States. A student does not manipulate the data, that is, does not change the details, as one does with spreadsheet programs, discussed later in this article.

Primary and intermediate grades

Figure 1 shows a partial passenger list of the ship *Oregon*. It can be used for problem solving and figuring statistical data.

Some problem solving questions and activities

- What was the age of the oldest passenger?
- What was the youngest age?

Students might be helped by the teacher's hinting, "An infant is usually someone under a year old. What would .09 and .08 mean?" (Nine months and eight months, respectively)

- Who had the most money? The least?
- Is it true that the older the men were, the more money they had? What information told you the answer?
- What other relationships can you see or think about from the ships record?

Young students will surprise you. They see much more that we think they see!

Some statistical activities

- What was the money range from least to most?
- What age occurs most frequently (the mode)?

Pressing the open-apple key and key A (for arrange) will sort data base information alphabetically (A-Z, Z-A) or numerically (0-9, 9-0) by the category on which the cursor rests. For instance, if you want all the males listed by age, the most recent sort is the final one. Therefore sort by age first and then by sex.

- What age occurs most frequently (the mode) among the males, females, children, adults, laborers, and so one?

Upper Grades

Some problem-solving questions and activities

- At which age did a person stop being a child and start to be an adult in the 1840s? (Age 10)
- If you wanted to have more money than the others, what three things seem to have helped? (Be male; be older—in the range 30-40 years; be a merchant)
- Only two females are shown as having money. Why were they the only ones listed?

The calculator can be used to find percentages; for instance, what percent of the people died, were male, and so on?

Some statistical questions

- Sort the information by age from oldest to youngest. At which person does the list split in half (exactly 13 people are above and below the person)? This approach is the first way to introduce the median-age concept.
- Sort from most to least money; find the monetary median. ($15.35)
- Using calculators, find the mean (arithmetic average) by age; by money. Is the mean or the median a fairer way to show the "average" money of the eleven people?

Even if the mean is figured on all twenty-three of

the survivors, it comes to $35.11, a less-than-fair picture of the actual monetary situation. Students are likely to identify with the financial status of the passengers, as shown in figure 1. The range of financial conditions of families in 1840 is close to the range of students' bank balances today. Many students receive an allowance close to the weekly pay of laborers in 1840!

Thought Strategies for Using Spreadsheet Information

Spreadsheet information uses formulas to calculate rapidly what would happen if data were manipulated under some fixed condition. The IMMIGRANT program presents budgets and markets as spreadsheet programs. Figure 2 shows a partial listing of the 1840 market price for foodstuffs. Column G shows how cost is figured.

Primary and intermediate grades

Ask young students to pretend that each of them has been given $10. Ask them to list vertically in column F the number of things they wish to buy in a week. Figure 3 shows how a typical second grade student might select food on the first try. Students look at the total and decide whether they are within their budget. Young students frequently do not understand the difference between ounces, pounds and bushels. It is a good idea for the teacher to have a cooking scale calibrated in ounces, empty pound and quart containers, and a bushel basket to demonstrate these units to students. Students need to decide if sixteen bushels of apples are needed for the Whelan family of nine people. Students can check decisions quickly by changing some of the nine figures in column F and see the resulting totals change simultaneously until they reach a week's total close to, but not over, $10. Students should be aware of

---

Figure 1   A Partial passenger list of the ship *Oregon*

```
File:   Oregon.partlist
Report: print.to.clipboard
```

Last name	First name	Age	Sex	Occupation	Money	Notes
Holland	Thomas	40	M	Merchant	185.00	
Weir	Charles	30	M	Merchant	525.00	
Whelan	Dennis	20	M	Laborer	25.45	
Whelan	Peggy	20	F	Housewife		
Whelan	Bridgett	13	F	Spinster		
Whelan	Catherine	11	F	None		
Whelan	James	28	M	Laborer	3.50	
Whelan	Catherine	28	F	Wife		
Whelan	Dennis	4	M	Child		
Whelan	Mary	3	F	Child		
Whelan	James	.09	M	Child		
Malehy	Bridgett	20	F	Spinster	2.50	
McWeeny	John	26	M	Laborer	2.68	
Leachman	Williams	40	M	Farmer		Died at sea
Leachman	Charlotte	40	F	Wife		Died at sea
Leachman	Elizabeth	12	F	Spinster		Died at sea
Leachman	Harriet	10	F	Spinster	15.35	
Leachman	Mary	8	F	Child		
Leachman	Richard	6	M	Child		
Leachman	Frederick	4	M	Child		Died at sea
Leachman	William	.08	M	Infant		
Leachman	Charlotte	.08	F	Infant		
Melegan	Thomas	20	M	Laborer	17.80	
Melegan	Sarah	20	F	Wife		
Marselly	Daniel	20	M	Laborer	1.50	
O'Donnell	James	30	M	Laborer	7.80	
McCorstand	Thomas	24	M	Mechanic	23.00	

Figure 2
An 1840 market listing for foodstuffs

```
File: Market.pl.form
          C    D      E      F      G
  4 Market Basket 1840
  5 Estimate your food costs for a single week.
  6 Order fractions of a unit when appropriate.
  7 Use APPLE-K to calculate totals.
  8 ----------------------------------------
  9 Item    Units  Price for  How    Cost
 10                each unit  many?
 11                                   F      G
 12 Allspice  oz     .03      E   0   (F12*E12)
 13 Apples    bu     .76          0   (F13*E13)
 14 Bacon     lb     .076         0   (F14*E14)
 15 Beans     qt     .1           0   (F15*E15)
 16 Beef      lb     .09          0   (F16*E16)
 17 Bread     loaf   .061         0   (F17*E17)
 18 Butter    lb     .21          0   (F18*E17)
 19 Candles   lb     .16          0   (F19*E19)
 20 Cheese    lb     .09          0   (F20*E20)


          Total for food
          (one week)         @ SUM(G12...G20)
          Total for food
          for year:            52*G22
```

how statistics shape their lives and influences their decisions. Students must decide what population is being studied in figure 3. It makes a difference if food is being bought for the Whelan family of nine people or the Malehy family of one person.

Upper grades

Figure 4 shows the food buys for Harriet Leachman as an upper grade student might choose them after all the budget items were figured and necessities were decided. The figures are based on a take-home pay of $7.50 a week from Harriet's job to keep her family of five together in one room of a rooming house when $2.50 was chosen as the limit for a week's worth of food.

Using data bases and spreadsheets with remedial-level students

The templates shown in this article are approximately one-fifth the size listed in the original program. A teacher may need to shorten templates even further, choosing records that help slower students see the relationships more clearly.

Using data bases and spreadsheets with gifted students

After working with IMMIGRANT, gifted students can create their own data base or spreadsheet for a new family. Perhaps students could update it for the situation facing today's immigrants—the Cuban boat people, the Vietnamese separated from family members, or migrant workers dealing with the green-card issue.

Did You Know?

1. More ideas for experimental teaching materials with IMMIGRANT have been developed by the Educational Technology Center of Harvard University, 337 Gutman Library, Appian Way, Cambridge, MA 02138. The original IMMIGRANT program can be purchased from the Educational Technology Center for $20.

Figure 3 A first attempt to spend $10. Note that the exact prices of bacon (.076) and bread (.061) are used in these calculations, but rounding to two places is done by the software for our convenience.

```
File: Market.pl.ex 1
          C    D      E      F      G
  4 Market Basket 1840
  5 Estimate your food costs for a single week.
  6 Order fractions of a unit when appropriate.
  7 Use APPLE-K to calculate totals.
  8 ----------------------------------------
  9 Item    Units  Price for  How    Cost
 10                each unit  many?
 11                                   F      G
 12 Allspice  oz    $.03   E   8   $   .24
 13 Apples    bu    $.76      16   $ 12.16
 14 Bacon     lb    $.076      2   $   .15
 15 Beans     qt    $.1        4   $   .40
 16 Beef      lb    $.09      10   $   .90
 17 Bread     loaf  $.061      3   $   .18
 18 Butter    lb    $.21       1   $   .21
 19 Candles   lb    $.16       6   $   .96
 20 Cheese    lb    $.09       7   $   .63
 21
 22            Total for food
 23            (one week)      $   2.48
 24            Total for food
 25            for year:       $128.70
```

2.   Catalogs listing public domain software are available from many sources for all types of computers; some even send a disk with new programs for a minimal price to cover materials and mailing. New Appleworks templates, from which to write your own problem-solving situations, are available through Public Domain Exchange, 2074C Walsh Avenue, Santa Clara, CA 95050.

3.   No significant differences were found in teaching these materials in two difference sequences—one beginning with applications of mean, median, and mode and one beginning with underlying concepts (Zawojewski 1986).

Have You Read?

Bitter, Gary G. *AppleWorks in the Classroom Today.* Watsonville, Calif.: McGraw-Hill/Mitchell, 1989.

Bitter, Gary G., Mary M. Hatfield, and Nancy Tanner Edwards. *Mathematics Methods for the Elementary and Middle School.* Needham Heights, Mass.: Allyn & Bacon, 1989.

Garfield, Joan and Andrew Ahlgren. "Difficulties in Learning Basic Concepts in Probability and Statistics: Implications for Research." *Journal for Research in Mathematics Education* 19 (January 1988): 44-63.

Morrison, Donald M., and Joseph Walters. "The Irish Immigrant Experience." Classroom Computer Learning 7 (October 1986): 40-42.

Strauss, Sidney, and Efraim Bichler. "The Development of Children's Concepts of the Aritmetic Average." *Journal for Research in Mathematics Education* 19 (January 1988): 64-80.

Wheeler, Fay. "The New Ready-made Databases: What They Offer Your Classroom." *Classroom Computer Learning* 7 (March 1987): 28-32.

Zawojewski, Judith Selman. "The Teaching and Learning Processes of Junior High School Students under Alternative Modes of Instruction in the Measures of Central Tendancy." *Dissertation Abstracts International* (December 1986): 47A 2068.

Figure 4
Purchases made from a weekly salary of $7.50

```
File: Market.pl.ex 2

Market Basket 1840
Estimate your food costs for a single week.
Order fractions of a unit when appropriate.
Use APPLE-K to calculate totals.
-----------------------------------------------
Item      Units     Price for  How     Cost
                    each unit  many?
```

Item	Units	Price for each unit	How many?	Cost
Allspice	oz	.03	8	$ .06
Apples	bu	.76	.5	$ .38
Bacon	lb	.076	2	$ .15
Beans	qt	.1	2	$ .20
Beef	lb	.09	5	$ .45
Bread	loaf	.061	3	$ .18
Butter	lb	.21	1	$ .21
Candles	lb	.16	3	$ .48
Cheese	lb	.09	4	$ .36

```
          Total for food
          (one week)        $  2.48
          Total for food
          for year:         $128.70
```

CHAPTER 4

Using Word Processors, Desktop Publishing Programs, and Graphics Packages in the Math Classroom

Although word processors, graphics packages, and related products such as desktop publishing programs are not as commonly used in the math classroom as spreadsheets and programming languages, they certainly have their uses. Mathematics involves words, numbers, symbols, and pictures, and all of these can readily be supplied and manipulated by such products. Word processors, desktop publishers, and graphics packages allow students to organize their thoughts and data and to represent concepts with pictures, charts, or graphs. With integrated packages such as AppleWorks GS or Microsoft Works, it is easy for students or teachers to produce professional-looking math papers. They make it possible for beginning computer users without artistic talent to produce the kind of document that, until recently, required highly expensive equipment (e.g.,typesetters) and highly skilled artisans, such as artists, editors, draftsmen, or layout specialists.

Word Processors

Writing should be an important component of the mathematics curriculum. The article by Davison and Pearce (1988) gives an overview of different ways in which writing may be used in the math classroom, from describing computational processes to exploring and conveying information related to mathematics (e.g., papers on famous mathematicians). Margaret I. Ford (1990), in her article, discusses the writing processes as it relates to problem solving. She draws parallels between the writing and problem-solving process and gives suggestions for helping children write their own problems to solve.

The word processor enhances the writing process by allowing the user to enter text into the computer, to view it on the computer's monitor, to make changes in the text, to format it as desired for printing, and to save it on disk for later use. Since the text is not printed on the printer immediately as it is typed, editing your work is much easier than with text typed on a typewriter. And, since text may be saved on disk, you can work on a document for a while, save it, turn off your computer, and reload it at a later date for further work.

Word processors have a number of features that make them powerful tools for manipulating text—far more powerful than a typewriter:

1. Word Wrap. When your text reaches the right edge of the screen as you are typing, your text "wraps"—that is, the entire word is moved to the beginning of the next line rather than being split in the middle. You do not need to press ENTER at the end of each line as you type. Indeed, you *should not* press the RETURN (or ENTER) key until you reach the end of a paragraph; otherwise, inserting and deleting text in the middle of a paragraph becomes more difficult (see Fig. 4.1).

2. Inserting Text. You may move the cursor (the blinking marker that indicates where the actions you take—such as entering text—will occur) into previously typed text using the arrow keys or mouse and *insert* new text. As you type, the text following the cursor will move over to make room for the new text. As the old text moves to and beyond the edge of the screen to accommodate the inserted text, it wraps to the next line (see Fig. 4.1).

Original Text

```
Although microcomputers became _available in the early
1970s, they were then still too expensive for schools
to purchase in any quantity.
```

Text with Insert

```
Although microcomputers became readily available in the
early 1970s, they were then still too expensive for schools
to purchase in any quantity.
```

Inserting with Carriage Returns at the End of Each Line

```
Although microcomputers became readily available in the
early
1970s, they were then still too expensive for schools to
purchase in any quantity.
```

Fig. 4.1: Inserting text in a word processor

The third example in Fig. 4.1 illustrates the problem that will occur if you press ENTER (or RETURN) at the end of each line as you do with a typewriter; since the text immediately following an ENTER is always left justified on a new line, a gap was created when "early" was moved to the next line to make room for "readily."

3. Deleting Text. You may delete text by moving the cursor to the right of the text you wish to delete and pressing the backspace (<-- key) or the DELETE key, depending on which brand of computer you use. The text following the characters you delete will move left and up to replace the deleted text. You may also delete blocks of text by first selecting the block, then pressing the DELETE button or selecting CUT from the appropriate menu.

4. Moving Text. You may select blocks of text and move them to new locations. First, select the text, then select CUT or MOVE from the appropriate menu, move the cursor to the new location, then select PASTE or press ENTER, depending on the word processor.

5. Copying Text. You may copy (duplicate) blocks of text using strategies similar to those for moving text.

6. Finding a Group of Characters. You may find a sequence of characters in your text. Typically, you select FIND from an appropriate menu in your word processor, enter the word you wish to find, then press ENTER. The computer will search through the text until it finds the string of characters and then place the cursor there.

7. Search and Replace. You may replace all occurrences of a particular string of characters in the text. Select REPLACE from the appropriate menu, then type the word you want to replace and the word you want to replace it with. Most word processors allow you to specify whether to automatically replace all occurrences of the word throughout the text or whether to stop at each word and allow you to decide whether it should be replaced.

8. Underline, Italics, Bold, Type Styles, Justification, Spacing. All word processors allow you to underline text or print it in italics or bold characters (if your printer has the appropriate capabilities). Macintosh-based and many of the new IBM-based word processors allow for a large variety of fonts, styles, and sizes. They will also allow you to right-justify or center your text; select the number of spaces printed between lines of text; or even choose the color of your text, if your printer has color printing capabilities.

9. Spelling Checker. An important feature of powerful word processors such as WordPerfect, Microsoft Works, and AppleWorks 3.0 is that they include a spelling checker. No teacher wants to be embarrassed by sending home or handing out papers with spelling or grammatical errors! Spelling checkers allow you to compare a word, group of words, or all the words in your document to the words in the program's dictionary. If a particular word does not match any of the words in the dictionary, the program will allow you to edit the word, skip it (e.g., names and technical vocabulary will not be in the dictionary), or substitute a word from a list of phonetically and orthographically similar words supplied by the word processor. See Fig. 4.2 for an example from WordPerfect.

```
^ A
Dear Mr. or Mrs. ^L,

    The fourth grade at Washington School is planning a field
trip to the beach, musium, and aquarium at Cabrillo National
Monument on Wednesday, March 1, 1992.  The bus will leave at
8:30 AM sharp and return by 4:00 PM.

    Please send a lunch and two drinks (cans of soda or boxes of
fruit juice) with your child on March 1. Please do not include
                                        Doc 1 Pg 1 Ln 1" Pos 1"
======================================================================

A. museum                B. mistime



Not Found: 1 Skip Once; 2 Skip; 3 Add; 4 Edit; 5 Look Up; 6 Ignore Numbers; 0
```

Fig. 4.2 WordPerfect Spelling

10. Thesaurus, Grammar Checker. Some, such as WordPerfect, contain a thesaurus; with the touch of a couple of buttons, the computer will print a list of the synonyms and antonyms of a particular word on the screen (see Fig. 4.3). There are even grammar checkers that are typically published separately from word processors. Once you have created your document with your word processor, you load the grammar checker and have it process the document. Grammatik (Reference Software), for example,

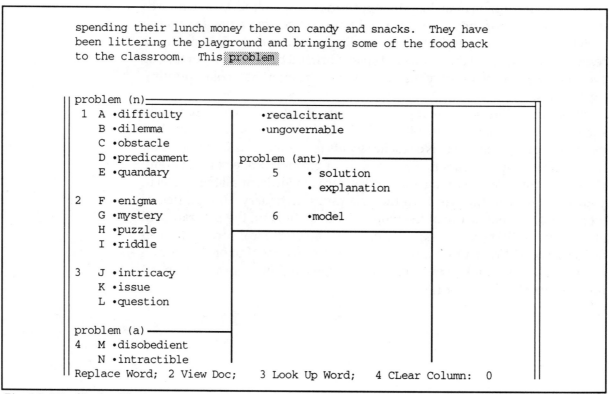

Fig. 4.3 WordPerfect Thesaurus

checks for thousands of errors in grammar, style, syntax, improper sentence construction, subject/verb agreement, passive voice, and spelling. Grammar checkers, however, will not help the writer in the important processes of the development of the theme, or paragraph or sentence construction.

11. Graphics. Many word processors, especially those designed to run on the Macintosh and newer IBMs, have built-in graphics capabilities. They allow you to draw lines, circles, rectangles, and the like, or to import graphics created on other packages into your document. The graphics capabilities of word processors are much more limited, however, than those of graphics packages (see below).

Desktop Publishers

The cousin of the word processor, the desktop publishing program offers many exciting features for the classroom. Desktop publishers function similarly to word processors—text wraps as you type it; you can insert, delete, and move blocks of text, etc; and so on—but have additional features that allow the user to design and print newsletters, signs, and posters, prepare overhead transparencies, and more. Typically, they allow you to print text in a large variety of sizes and styles, to print in columns, and to create graphics.

Many desktop publishers and graphics packages provide large sets of pre-made graphics. Programs such as Show Off, Print Shop, News Room, and Publish It! offer many dozens of pictures of all sorts. There are even third-party publishers who sell diskettes full of pre-made graphics that can be used with desktop publishing, graphics, and even word processing programs.

A good desktop publishing program (e.g., Publish It!, First Publisher, Page Maker) will have many or all of the features mentioned above—a large variety of type styles and sizes (fonts); the ability to easily arrange words, symbols and numbers on the page; the ability to draw lines and shapes and shade them as desired; the ability to create your own graphics; color; and the presence of a large variety of pre-made graphics. Of course, you should be able to use all the desired features with your computer and printer. Fig. 4.4 shows a math worksheet created with PageMaker, a powerful desktop publisher for Macintosh or IBM computers.

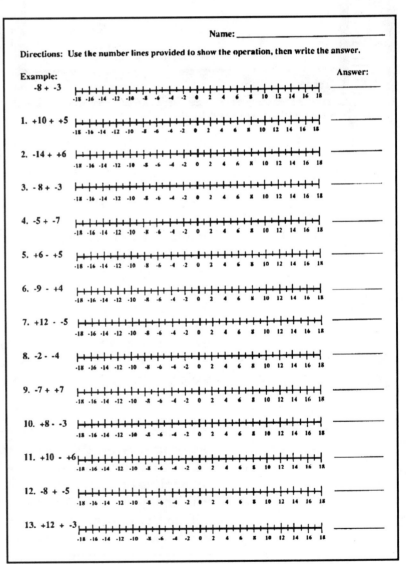

Fig. 4.4: Worksheet created with PageMaker.

Graphics Packages

Although many word processors and desktop publishers have graphics capabilities, stand-alone graphics packages tend to have more features. In general, they have tools that allow you to draw squares and rectangles; circles, ovals, and arcs; diagonal lines; vertical and horizontal lines; and freehand lines; to erase or move sections of a drawing; to fill shapes in a

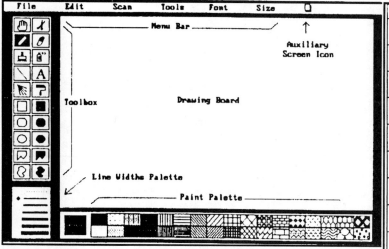

Tool Icon	Name	Function
	Scroll	Move picture or selection.
	Selection	Define an area to edit.
	Pencil	Draw freehand lines. Magnify area to the pixel or dot level.
	Adjustable Eraser	Erase areas of picture under cursor. Clear picture from the drawing board.
	Paint Brush	Paint picture with brush strokes. Change brush cursor.
	Spray Paint	Spray paint on or shade picture. Change spray width.
	Line	Draw straight lines.
	Text	Type text on picture.
	Lines Fill	Draw graphic line designs.
	Area Fill	Color a defined area.
	Rectangle	Draw a hollow or filled rectangle or square.
	Rounded Rectangle	Draw a hollow or filled rectangle or square with rounded corners.
	Oval	Draw a hollow or filled oval or circle.
	Polygon	Draw a hollow or filled polygon.
	Freeform	Draw a hollow or filled freeform shape.

Fig. 4.5: Paintshow Plus screen and Toolbox.

desired shading or color; or to manipulate the shapes on the screen in a variety of other ways. Graphics packages also allow you to print words and numbers. However, the emphasis is on the graphics, not the text. Often they have very few choices when it comes to font styles and size. Furthermore, they treat text as graphics—you position the text on the page and define it. And once created, text is not as easy to modify as it is with word processors or desktop publishers.

The various graphics tools in a graphics package are selected from a toolbox using a mouse[1]. To select a shape, point to the icon that represents the shape and click the mouse button. Next, point to the drawing area of the screen and define the shape. For example, to draw a blue rectangle with Paintshow Plus (an IBM-based graphics package by Logitech), first select blue from the Paint Palette, select the filled rectangle from the Toolbox, then position the cursor on the point where you wish the upper left corner of the rectangle to be, press the mouse button and drag it to the point where you want the lower right corner to be and release the button. The blue rectangle will now be drawn. Fig. 4.5 shows the drawing screen (with Toolbox and Paint Palette) and an explanation of the tools available in Paintshow Plus.

---

[1] Most graphics packages are designed to be used with a mouse. Although some graphics packages allow you to use the keyboard instead of a mouse to select tools, to draw, and so on, it is much easier to use mouse-based graphics systems.

Student Use of Word Processors, Desktop Publishers, and Graphics Packages

Word processors, desktop publishers, and graphics are of great value to students in the areas of problem solving and mathematical applications. They can also be useful in activities designed to develop understanding of mathematical concepts.

A word processor provides a convenient way to present problems and a problem-solving schema to students. The problems can be entered by the teacher into the word processor, the problem-solving steps listed, and (possibly) a few hints included. The student then types the solution steps and, eventually, the answer at the computer. The resulting document (with the problem, solution steps, and answer) can be saved and either a printout or the data disk can be used by the teacher to evaluate the student's responses.

Another good use of word processors is to allow students to write their own problems. The teacher can, first of all, establish some pa-

Fig. 4.6: Students Working with a Word Processor on a Problem Solving Task

rameters. For example, you might want to post a list of rules regarding the problems to be written (e.g., "The problems must involve addition of more than three whole numbers, and the problems must contain extra information"). Next, have students work at the computer, perhaps in groups of two, three, or four, to create the problems. The teacher then checks the problems for appropriateness before giving them back to the students to solve. The students could work at the computer and enter the strategies they use as they attempt to solve the problems and, finally, their answers.

A variation on the above activity is to create a set of problems on the word processor where the result or answer is given, but the values in the problem or part of the description of the problem are either incorrect or missing. The students, then, would enter the missing (or incorrect) information, using the editing features of the word processor. With certain problems, it might be helpful to also have a spreadsheet available. For example, if the student is working with AppleWorks or Microsoft Works, he/she could move to that application to do the computations, then move back to the word processor (possibly moving any tables or graphs created from the spreadsheet to the word processor).

Pupil Activity 4.1: Word Processor Problem Solving

Objective
Solve problems and enter results on word processor.
Grouping
One or two pupils.
Materials
Computer, word processor, problems.
Procedure
Students are given a set of problems like the example below on disk. They must enter missing text and print out completed problem.
Sample Problem
Giant Department store is having a sale. The regular price of a radio is $32.00._____. The sales price is $24.00. (Add a sentence in the blank spot to complete this problem.)

Pupil Activity 4.2: Problem Solving with Integrated Application (Spreadsheet and Word Processor)

Objective
Solve problem on spreadsheet, move spreadsheet to word processor, print results.
Grouping
One or two pupils.
Materials
Integrated package.
Procedure
Student given the problem below with instructions in a file (created with word processor).
Sample Problem
At the Stamp Collector's Swap Meet, John bought several sets of old stamps from different countries. The sets contained the following stamps: 6 U.S. stamps costing _____ each, 8 Canadian stamps costing _____ each, 21 Mexican stamps at _____ each, 14 French stamps at _____, and 7 Cuban stamps at _____. If the total cost of the stamps was exactly $10.00 and single stamps from the different countries cost 17, 10, 21, 12, and 50 cents (i.e., the stamps from, say, France cost 17 cents apiece—the prices are not in order), what was the cost of single stamps from each country?

   Type your answers here and print out the results, following these instructions:
1. Restate the problem in your own words.
2. List everything you know about the problem.
3. Set up a spreadsheet like the one below. Experiment with the stamp prices until the prices are correctly matched with the countries (the total should be $10.00—there should be a *formula* at cell D7 that finds the total of Column D).
4. Copy the completed spreadsheet back into the word processor, enter the missing values in the problem above, and print out the whole thing, including the problem and the spreadsheet.

	A	B	C	D
1	Country	Stamp Price	No. Stamps	Total/Country
2	U.S.A.		6	
3	Canada		8	
4	Mexico		21	
5	France	$.17(?)	14	$2.38(?)
6	Cuba		7	
7			Total	
8				
9			Total Cost	$10.00

Fig. 4.7: Problem on integrated package.

Another place where word processors and desktop publishing programs are useful is student reports on data-gathering/statistics activities. For example, students could gather data on the pets owned by children in the class (e.g., length in centimeters, weight in kilograms, number of grams of food eaten a day, amount of water drunk in a day). Other tools (e.g., spreadsheet or calculator) could be used to do the data analysis (e.g., calculate averages). The word processor could then be used to report the results.

Desktop publishing packages or word processors could also be used to create electronic worksheets for practicing mathematical concepts or operations. The worksheets would be saved on disk and the students would answer the questions on-screen and save or print out the results. Word processors or desktop publishers with good graphics capabilities could even be used by students to represent problems graphically. Fig. 4.8 shows part of a screen-based worksheet designed to help students understand the meaning of addition of fractions, before students learn the algorithm for adding fractions.

Pupil Activity 4.3: Representing Fractions with a Graphics Program

Objective
Draw pictures using graphics package to represent the sum of fractions.
Grouping
Individuals.
Materials
Graphics package with problems saved on disk.
Procedure
After students have learned to use the graphics package, give them problems like those in the example.
 Alternative: Give students the problem on a worksheet and have students solve the problem using the graphics package and print out the solutions.

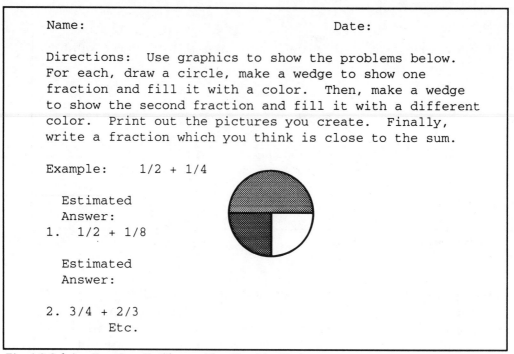

```
Name:                                    Date:

Directions:  Use graphics to show the problems below.
For each, draw a circle, make a wedge to show one
fraction and fill it with a color.  Then, make a wedge
to show the second fraction and fill it with a different
color.  Print out the pictures you create.  Finally,
write a fraction which you think is close to the sum.

Example:     1/2 + 1/4

  Estimated
  Answer:
1.  1/2 + 1/8

  Estimated
  Answer:

2. 3/4 + 2/3
       Etc.
```

Fig. 4.8 Solving Fractions Problems with a Graphics Package

Pupil Activity 4.4: Numeration with a Graphics Package

Objectives
Represent numerals using pictures created with a graphics package.
Grouping
Individuals.
Materials
Worksheet; graphics package.
Procedure
Students are given a set of numerals to represent graphically (either on disk or on paper). Students use graphics package to draw pictures to represent the numbers.

```
Use the mouse to draw squares and rectangles to show the
following numbers:

Example:     357

1. 526


2. 607
```

Fig. 4.9 Drawing shapes to represent numbers

Graphics packages or the graphics component of word processors or desktop publishers may be very useful in helping students solve verbally-stated problems involving pictures or models. Problem Example 3 in Chapter 1 (See Fig 1.13) is an example. Another example follows:

Ashe Street runs East and West. Alice, Beth, Carl, and Doris all live in different houses on the same side of Ashe street. Their house numbers are 28, 32, 34, and 36, but not necessarily in that order. Alice and Carl live to the East of Doris. Beth's house number is higher than Doris'. Carl's house number is the lowest. What are the street numbers of each person's house?

Students could use the graphics package to draw a picture to help them solve the problem—two lines to represent the street, and numbered boxes to represent the houses. If your graphics package allows you to identify and move graphics units (e.g., dragging with a mouse), you could place names in the boxes to represent people's houses and place them in the desired location to solve the problem. (See Figure 4.10.)

Mathematics, like all other parts of the curriculum, involves writing. Word processors can be of great value to a student in the composing and editing process, especially in the problem-solving process. Not only are desktop publishers and graphics tools for exploring shapes and relationships, but they allow students to produce high-quality documents. In turn, students' self-concept in the mathematics classroom may be enhanced.

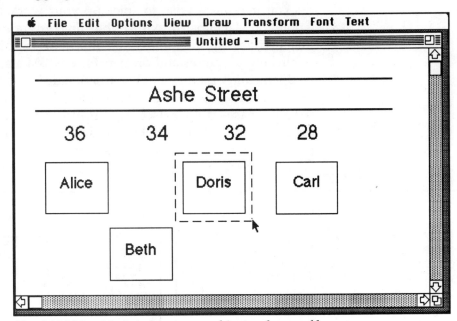

Fig. 4.10 Using Macintosh screen graphics to solve a problem.

Teacher Use of Word Processors, Desktop Publishers, and Graphics Packages

The major use of word processors, desktop publishers, and graphics programs by teachers is to prepare student materials. Tests, worksheets, and other handouts created on a word processor and saved to disk can be quickly and easily modified and printed for each subsequent use. Prior to the word processor, deleting several questions in the middle of a previously prepared test meant retyping the whole document, or cutting, pasting, and using white-out. Now, all that is needed is to use the block delete feature of the word processor, then to move the cursor down and change the numbering of subsequent questions. Many other features of word processors assist the teacher in preparing student worksheets. Tabs column settings help format pages, underline and bold (as well as alternative fonts) help draw the student's attention to important material in the text. Block move and copy features help in the editing process. They also allow you to pull pieces of text from several documents and put them together to form a new document.

Lower-grade teachers can benefit greatly from desktop publishers or word processors such as Microsoft Works or AppleWorks GS that have desktop publishing features. For example, they can design problem-solving or computational worksheets using large fonts appropriate for beginning readers (see Fig. 4.11).

Lines, boxes, and the like, can be quickly and easily placed on the worksheet to organize students' steps in the problem-solving process (Fig. 4.12). Or, you could create problem-solving worksheets that substitute pictures for words (see Fig. 4.13).

You may even wish to spice up your worksheets with color—many Apple and Macintosh desktop publishing programs have this capability and Imagewriter II printers have the ability to print in color. Many desktop publishing programs and printers for the IBM also have this capability.

Elementary mathematics instruction depends heavily on manipulative materials (models). For example, colored chips are used to teach number concepts; base-10 blocks (see Fig. 4.14) are used to teach our number system, operations on whole numbers and decimals, and metric measurement; place-value charts are used to teach whole-number operations; number lines are used to teach about our number system and beginning operations on

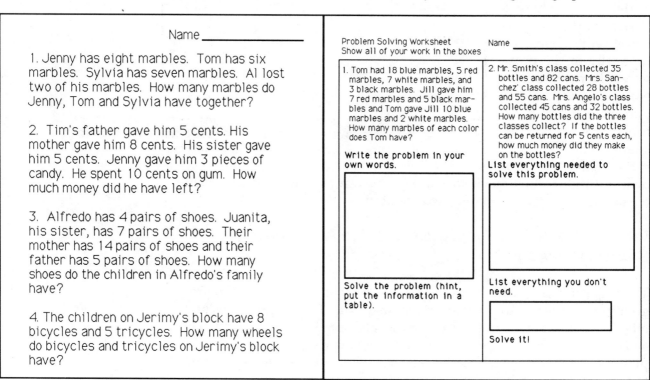

Fig. 4.11 Worksheet with large fonts

Fig. 4.12 Problem-solving worksheet with boxes, Microsoft Works

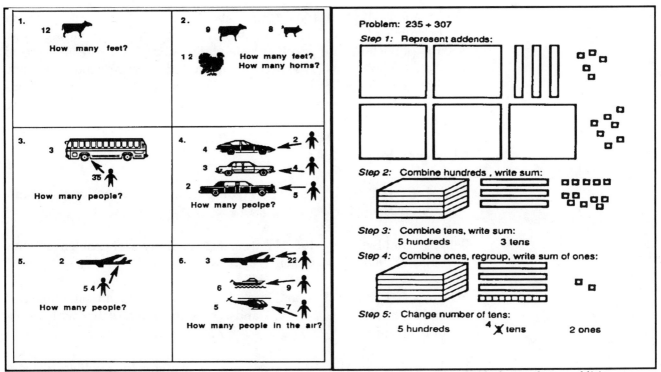

Fig. 4.13: Problem-solving worksheet with pictures

Fig. 4.14: Using base-10 blocks to solve an addition problem (from Wiebe, 1988, p. 153)

integers; and so on. Usually, a child's first experience with a mathematical concept (e.g., multiplication of one-digit numbers) will be with the real objects or with a model. Pictures are later substituted for manipulatives because manipulatives are difficult to manage in the classroom (for example, how do you evaluate the responses of a class full of pupils using colored chips to do subtraction problems?). Desktop publishers or word processors with graphics capabilities allow you to create math worksheets that involve pictures of manipulative materials or other objects.

Teachers, like most personal computer users, will find that computer tools that allow them to create, manipulate, and print text, symbols and graphics—word processors, desktop publishers, and graphics packages—are extremely useful. The capability of these tools to place graphics on the screen or on paper can help the teacher make mathematics more meaningful. And, students will find these tools to be useful in solving and representing a variety of mathematical problems and situations. If you are setting up a laboratory or purchasing software for use in your

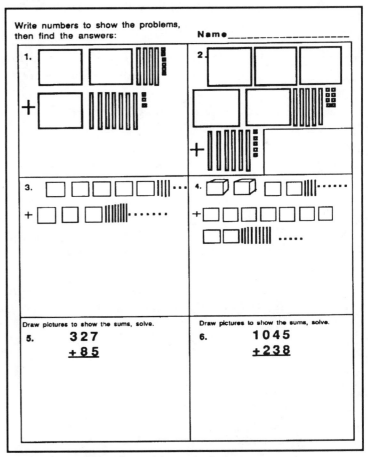

Fig. 4.15: Math worskeet with manipulatives

and situations. If you are setting up a laboratory or purchasing software for use in your classroom, consider purchasing an integrated package with desktop publishing and graphics capabilities, along with spreadsheet and data base subprograms. The cost of an integrated package is much less than purchasing each application separately, and you will have the benefits of a common set of commands for all the subprograms and of being able to move data from one application to another.

Bibliography

Abel, J. P., and Abel, F. J. 1988. Writing in the mathematics classroom. *Clearing House* 62 (December): 155-58.

Bell, E. S. and Bell, R. N. 1985. Writing and mathematical problem solving. *School Science and Mathematics* 85(3, March): 210-21.

Edwards, N.T., Bitter, G.G., and Hatfield, M.M. 1990. Teaching mathematics and technology: Data base and spreadsheet templates with public domain software. *Arithmetic Teacher* 37 (8, April): 52-55.

Ford, M. I. 1990. The writing process: A strategy for problem solvers. *Arithmetic Teacher.* 38(3, November): 35-38.

Johnson, M. L. 1983. Writing in mathematics classes: A valuable tool for learning. *Mathematics Teacher.* 76(2, February): 117-19.

Wiebe, J.H. 1992. Wordprocessing, desktop publishing, and graphics in the mathematics classroom. *The Computing Teacher* 19(5, February) 39-40

Discussion Questions

1. How do word processors differ from desktop publishing programs? How are they similar? How do graphics packages differ from desktop publishing programs? How are they similar?

2. Students could do worksheets created with a word processor at the computer or the worksheets could be printed out and students could complete them at their desks. What are the advantages and disadvantages of screen-based vs. paper-based worksheets?

3. Can you think of some possible disadvantages of integrated packages over packages containing single applications?

4. Discuss the uses of writing and pictures in the mathematics classroom.

Activities

Activity 4.1

Objective
Use a desktop publisher to create a mathematics worksheet.
Rationale
Use of professional tools such as a desktop publisher can add a feeling of professionalism to the work of a teacher. If properly done, it can add motivation to your students and/or help make mathematical concepts meaningful.
Materials
Desktop publisher, computer, printer.
Advance Preparation
Read this chapter. Learn to use the desktop publisher or the desktop publishing features of your word processor.
Procedure
Use a desktop publishing program to create a mathematics worksheet containing graphics for children at the grade level you teach or intend to teach. The graphics either should be used to help motivate students' interest in the worksheet or should be pictures designed to help make mathematics concepts meaningful, as in Fig. 4.15.

Activity 4.2

Objective
Use an integrated package to solve a problem in math.
Rationale
The ability to manipulate data in one application of an integrated package and move it to another application for further manipulation or creating a report is a powerful tool for classroom management or problem solving.
Materials
Integrated package, computer, printer.
Advance Preparation
Learn to use integrated features of your applications package.
Procedure
Use an integrated package (e.g., AppleWorks or Microsoft Works) to solve the problem in Pupil Activity 4.2 above.

Using Writing Activities to Reinforce Mathematics Instruction

by David M. Davison and Daniel L. Pearce

The traditional view has been that students learn to write in English classes and to compute in mathematics classes and "never the twain shall meet." Certainly little thought has been given to the idea that teachers of various subject areas, especially mathematics, should seek to have students engage in writing activities as part of their study of that area. In recent years, however, this position has been changing, and different authors have recommended increased writing about mathematics by students as a useful and valuable aspect of mathematics instruction (Burton 1985: Greenius 1983; Johnson 1983; Shaw 1983; Watson 1980).

One of the ways students acquire new information

their students. Performing a writing task requires students to reflect on, analyze, and synthesize the material being studied in a thoughtful and precise way.

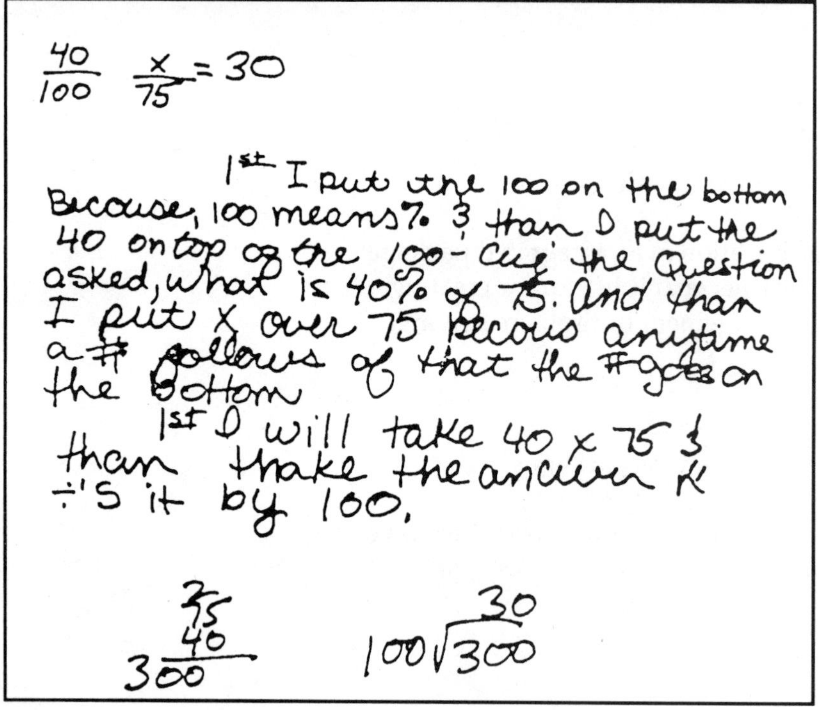

The authors have been exploring various ways that teachers use writing in mathematics instruction. The results of these investigations have suggested that the use of writing activities is sporadic in junior high school mathematics classes (Pearce and Davison, in press). However, a pattern of student writing activities has emerged. In particular, it appears that writing activities used in mathematics classrooms can be classified into five categories: direct use of language: linguistic translation; summarizing; applied use of language: and creative use of language. Writing activities in each of these categories have a use in the mathematics classroom. The purpose in this article is to present different activities for each of these categories and to encourage mathematics teachers to try implementing these suggestions in their own instruction.

Direct Use of Language

This level of writing involves students in copying and recording information. Although most mathematics teachers require that students take notes, notetaking is more effective if the students are given some structure. Many of the teachers checked their students' notebooks periodically, but the students viewed the notebook as more important when the checking was more frequent. In one classroom, the teacher would spot-check the students' notebooks daily and give the

Figure 1 A student's efforts to describe a solution to 40/100 =x/75

is through putting ideas into language. Both Emig (1977) and Haley-James (1982) have stressed that writing is a mode of language that particularly lends itself to the acquisition of new knowledge. Writing about something involves many of the thought processes mathematics teachers would like to foster in

class advice on maintaining the notebooks as a useful study tool. Another teacher impressed on the students the necessity for completeness and accuracy in their notebooks by giving periodic quizzes in which the students had to use their notebooks to find the answers.

Linguistic Translation

Activities in this category call for the translation of mathematical symbols into written language.

One task in this category is the translation of symbolic expressions into words. The following is an example of this type:

Write in words: (2n +5)/3

Here the student makes a straightforward translation from the symbolic meaning to the English language. Variations of this exercise would include the following:

Explain the meanings of the symbols in the formula $d = rt$.

Translate the formula $A = lw$ into words.

Another important task in this category is having students translate their solution to a word problem into a complete English sentence. For example. the solution may be $x = 9$, but students need to complete the problem by writing "The length of the garden plot is 9 meters."

Writing activities within this category do not necessarily have to be rigid exercises in translation. One teacher gave the class a long division problem:

$$23 \overline{)276}$$

Each student wrote out the problem in English and then wrote out the steps to solve the problem. The teacher then conducted a whole-class discussion in which various students read their written responses aloud. These were written on the chalkboard. Then the teacher showed an overhead transparency of his writing of this problem. Students were allowed to compare their answers to that of the teacher. This modeling was important because it allowed the students to know what was expected of them.

As a follow-up exercise, each student repeated this writing process, working with an individually designed problem. This written problem was given to another student, who had to reconstruct and solve the problem from the written sheet and give the results in mathematical symbols. The students found this activity enjoyable and approached it as a puzzle. This procedure varied the routine of giving students whole-class assignments.

Summarizing

Summarizing activities include paraphrasing or summarizing material from the textbook or some other source (such as classroom presentations).

Writing activities in this category can take a wide variety of guises. For example, rather than just present the material for transcription into notes, the teacher could discuss the material and then ask the students to record notes from memory. One extension of this type of activity is to have the students describe solutions in detail. One such example would be to have the students describe how they solve proportion problems (fig. 1). Another example, a student's description of how to multiply two fractions, is shown in Figure 2. Writing of this type can either be incorporated into existing notebooks or be recorded in a special-purpose notebook.

Another summarizing activity is to have students keep a personal mathematics journal. A journal can consist of reactions to the mathematics being studied, such as questions about specific problems encountered and personal insights into the processes of mathematics. One such example is "Great! I found a new way to change decimals to fractions." When the teacher read the journal it showed her how successful the student was in understanding the material. Simi-

Figure 2 A student's explanation of how to evaluate 2 2/3 x 9/10

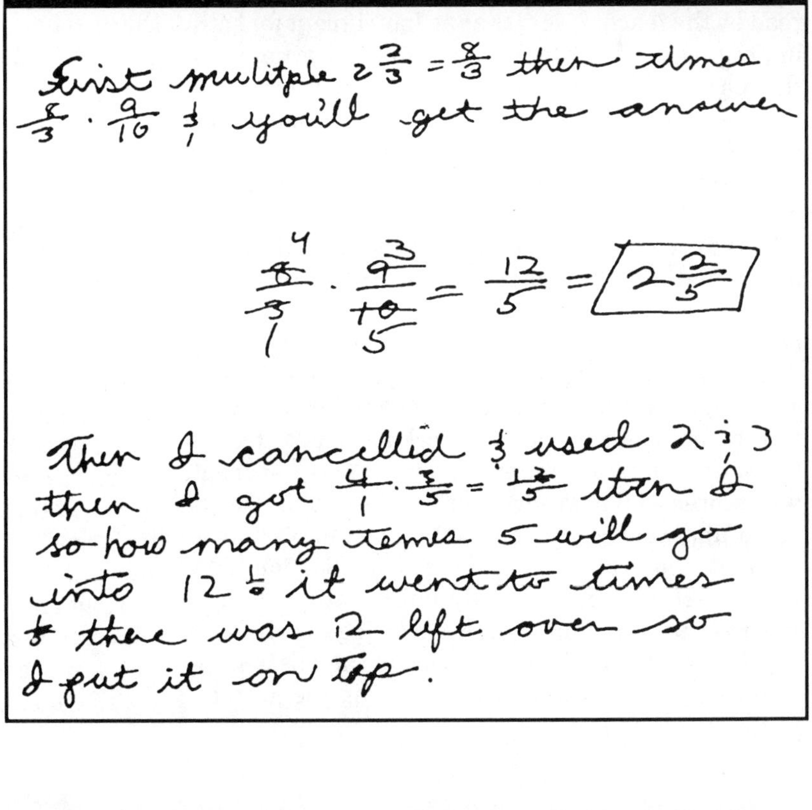
First mulitple $2\frac{2}{3} = \frac{8}{3}$ then times $\frac{8}{3} \cdot \frac{9}{10}$ & you'll get the answer

$\frac{8}{3} \cdot \frac{9}{10} = \frac{12}{5} = 2\frac{2}{5}$

Then I cancelled & used 2 ; 3 then I got $\frac{4}{1} \cdot \frac{3}{5} = \frac{12}{5}$ then I so how many times 5 will go into 12 ½ it went to times & there was 2 left over so I put it on Top.

Figure 3 A picture of a car prompted a student to create this example

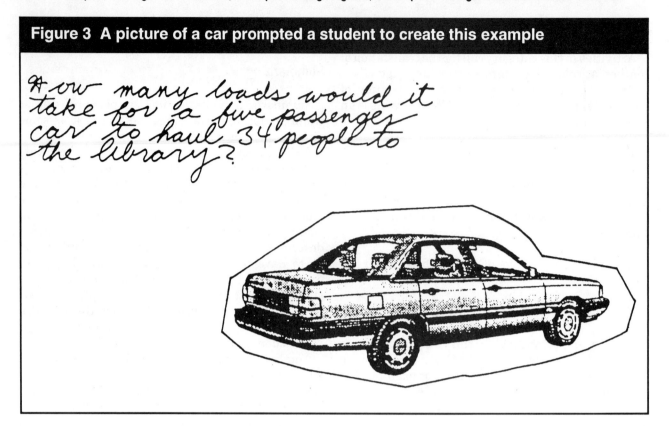

How many loads would it take for a five passenger car to haul 34 people to the library?

larly, the journals of other students pointed up strengths and weaknesses in their comprehension of the material.

It should be noted this is the first writing application discussed thus far that is not intended specifically to improve mathematics performance. A journal is a communication device and certainly, when read by the teacher, serves as an instrument for both improving writing ability and communicating with the teacher. If attention is paid to the content of the journal, in contrast to mechanics of language (spelling, sentence fragments, punctuation), the teacher will learn how the student is interacting with the mathematics being studied.

This form of writing can help students learn more about mathematics in two ways. First, the act of writing about the material serves to clarify the mathematical process in the student's mind. Thus, students communicate how well they understand the content. A student may have misunderstandings that do not show up in homework assignments and tests. Such misconceptions can trouble the student who scores well in graded work but is unclear about some apparently minor detail. An example is the student who wrote, "I'm glad I got that question right on the test, but I'm not really sure what I'm doing." Because no warning signals are triggered by test results, such a student may believe the misconception to be unimportant. By examining the journal,

the teacher can be alerted to potential or actual problem areas and take appropriate remedial action.

Summarizing activities involve the students in such important processes as considering and reflecting on mathematical concepts and using mathematical terms. To be most effective, however, students summarizing efforts need to be read periodically by the teacher to insure accuracy of content and to provide feedback. However, this feedback does not always have to come from the teacher. It can also come from other students who read and comment on each other's work.

Applied Use of Language

The applied use of writing activities includes situations where the student applies a mathematical idea to a problem context.

In many classrooms, the students were required to make up story problems. In one instance, the teacher had the students cut out a picture from a newspaper or magazine and write a two-step story problem based on the picture (figure 3). In another instance, the class had just completed a unit on solving two-step linear equations in one variable. As an exercise, the students were given a set of equations and were asked to write related story problems. By way of follow-up, each student was assigned the task of making up another story problem that would lead to a two-step linear equation. After the problems were written, students exchanged problems with their neighbors, who solved the problems and returned them for correction. The teacher collected the original problems and their solution efforts for grading purposes.

Another teacher indicated that she asked each student to write one problem related to the topic just completed. From the problems submitted, she selected a set to give to the students as homework. The teacher also retained some questions for future use on a quiz or test.

A different approach used by another teacher was having students prepare quiz or test questions. The students turned in their questions and solutions as a homework exercise. The teacher then selected some of these questions for inclusion on the unit test.

Activities of this type are important because the ability to write about mathematics shows that the student can apply the concepts learned to a real-world situation. In particular, having students write their own story problems is an effective way to have them analyze problem-solving strategies. For example, by exchanging problems with other students, each student gets practice in solving problems that are more realistic than typical textbook exercises. Students are also more inclined to make an honest effort to solve problems that are viewed as more realistic.

Creative Use of Language

In the category of creative use of language, the student uses written language to explore and convey information that, although it may contain mathematics concepts, is not specifically being studied in mathematics.

Potential assignments in this category typically are viewed as outside the development of mathematical skills. Although many of the teachers we have worked with agreed that such assignments allow students the opportunity to apply mathematics skills to real-world situations, relatively few of them use such assignments. One teacher assigned the writing of short papers on important people and events in the history of mathematics. As an alternative, the students could make oral reports to the class and submit their speaking notes to the teacher. Sample topics included "ancient attempts to find the value of pi" and "Pythagoras's contribution to the study of numbers."

Research projects represent a significant means of enabling students to integrate writing and mathematical skills. More than one teacher used statistical projects in this way. Groups of students were assigned to investigate a topic and to write a report.

One example was an investigation into whether students at different grade levels were more (or less) likely to watch situation comedies on television. The data-collection process gave students an opportunity to experience what happens in this type of project. In writing the report the students synthesized the findings of the research and communicated them in a coherent narrative. Later, the teacher posted the papers on a bulletin board for the class to read.

Summary

Having students engage in writing tasks in a mathematics context has several benefits. Not only will the opportunity to practice writing improve a student's ability in written expression, but using writing to practice mathematical tasks will also assist students in comprehending mathematics concepts and improve their ability to communicate mathematically. The inclusion of writing activities, such as those presented in this article, potentially has additional value when the kinds of activities are varied and writing is treated as an attempt at communication. In such an environment writing can become both a personal and a rewarding activity for students. Given that many students view mathematics as a stringent program of rules, facts, and figures, writing activities can involve students in useful and enjoyable mathematical activities. These, in turn, can encourage students to become more proficient in mathematics.

References

Burton, Grace M. "Writing as a Way of Knowing in a Mathematics Education Class." *Arithmetic Teacher* 33 (December 1985):40-45

Emig, Janet. "Writing as a Mode of Learning." *College Composition and Communications* 28 (May 1977):122-28.

Greenius, Eric A. "Sharing Teaching Ideas: Notebook Evaluations Made Easy!" *Mathematics Teacher* 76 (February 1983):106-7.

Haley-James, Shirley. "Helping Students Learn Through Writing." *Language Arts* 59 (October 1928):726-31.

Johnson, Marvin. "Writing in Mathematics Classes: A Valuable Tool for Learning." *Mathematics Teacher* 76 (October 1983):117-19.

Pearce, Daniel L. and David M. Davison. "Teacher Use of Writing in the Junior High Mathematics Classroom." *School Science and Mathematics* (in press).

Shaw, Joan G. "Mathematics Students Have a Right to Write." *Arithmetic Teacher* 30 (December 1983):16-18.

Watson, Margaret. "Writing Has a Place in Mathematics Class." *Mathematics Teacher* 73 (October 1980): 518-19.

The Writing Process: A Strategy for Problem Solving

by Margaret I. Ford

Over the past decade, mathematics educators have promoted problem solving as the goal of school mathematics. Yet in 1987, the National Assessment of Educational Progress revealed that our nation's schoolchildren are still falling short of our goals for their problem-solving abilities. Many students dislike word problems in mathematics, and many teachers report feeling frustration and discouragement in helping their students learn how to solve such problems (Ford 1988). What can teachers do to improve students' attitudes toward problem solving and to realize the goal of helping students become better problem solvers?

The new *Curriculum and Evaluation Standards for School Mathematics (Standards)* (NCTM 1989) suggests that mathematics should be communicated in writing because writing in mathematics promotes meaningful learning. Writing can be used in several ways as a strategy to make the process or activity of problem solving meaningful to students. Perhaps part of the solution to helping students become successful problem solvers can be found in a strategy borrowed from the language arts. Let us examine this strategy for dealing with one aspect of problem solving, that is, the solution of routine story problems.

Writing and Problem Solving

Studies of students' difficulties in learning to write reveal that simply asking students to write, or giving them opportunities to write, does not produce better writers. Researchers have found that guiding students through a writing process is the key to producing better writers and better thinkers (Graves 1983; Calkins 1986). The task of developing problem-solving ability is similar because problem solving, like writing, is a process. Simply asking students to solve problems, or presenting opportunities for students to solve problems, does not produce better problem solvers. Guiding students through the processes of problem solving is essential if they are to become better problem solvers.

Process-focused problem-solving lessons address the aspects of problem solving that are typically most difficult for students. Suydam (1980) indicates that students have difficulty with problems that have a lot of words and complex sentence structures and with problems in which information is not presented in the order in which it is used to solve the problem. Fennell and Ammon (1985) suggest that asking students to write their own word problems is an effective strategy for the teaching of problem solving. Teachers can help students to become better thinkers by using the students' own language in developing problems to be solved.

The Writing Process

Although many teachers suggest that students write problems as part of instruction in problem solving, this strategy seems absurd for some students. How can students learn about something as difficult as problem solving by doing something equally as difficult (writing stories)? The key is to use the same strategies that are effective in process-focused writing instruction.

The basic components of the writing process include five stages. In the "prewriting" stage the teacher furnishes a stimulus for the writing and helps students to think about what they already know. In the "writing" stage students simply write down their thoughts and ideas without the pressure to produce a perfect clean copy of their writing. In the "conference" stage the teacher or peer listens to what the writer is saying and gives feedback by asking specific questions that focus on the content of what has been written. In the "revision" state students focus their ideas, clarify their writing, and add or delete information. They also learn to see their writing through someone else's eyes. In the "publication" stage students put their work into a final form for others to read.

The Writing Process in Problem Solving

An amazing result of applying the writing process to problem solving is the positive attitude of the students as they engage in the process. The changing of attitudes is only the first step toward the development of problem-solving ability, but it is a crucial one. The following description of two mathematics classrooms that use the writing process to write and solve routine story problems illustrates the benefits of this approach to instruction.

The first example is a class of third graders who

have had an indifferent attitude toward problem solving. As often happens, even those students who are good at computation have had some difficulty in solving word problems. Most of their problem solving has involved typical story problems from their textbook. I had no difficulty in getting the attention of the students because for the most part third graders are eager to learn. I began the session with an overview of the writing process and an explanation of the stages we would be going through in the class. I did not ask the students to write a particular type of word problem, such as one involving addition or multiplication, one or two steps, but instead moved right into the prewriting stage of the writing process.

Prewriting

I gave menus and catalogs from local businesses to the students and asked them to think about a situation in which they might be dining at one of the restaurants or shopping at one of the stores. I gave them the following example:

> My sister and I went to lunch. I was very hungry, so I ordered fried chicken, which cost $3.99. My sister ordered a salad for $1.99. We each ordered a glass of iced tea, which was $0.50 a glass. If we had $7.00, could I also buy a dessert for $1.00?

The students were not asked to solve my problem but rather to think about situations in which they might be involved that were similar to mine. My story problem was rather complex for third graders. Two-step problems are often not introduced until fifth grade, and my example involved more than two steps. This example concluded the prewriting stage, and the students were asked to write a word problem.

Writing

The students began the writing stage by writing out their own ideas on yellow ledger paper. The yellow paper served as a visual reminder that they were not writing a finished copy. The students also had calculators available to work out their problems. My primary interest was not in their computational ability but in their ability to solve the problems. Having a calculator removed any misgivings the students might have had about performing the calculations necessary to solve their problems. All around the room students were busily looking through their menus and catalogs and writing down their ideas. When the students finished writing, they read over their own work and performed their calculations before they moved into the next stage of the writing process: the conference.

Conference

During the conference stage students moved into pairs and read each other's work. In this particular session I did not model a conference for the students. I asked them to read each other's problems to see if they made sense and to work out the problems using calculators to see if the problems were workable. They were also reminded to discuss with each other what they thought about the problems.

In groups of two, students solved each other's problems to make sure that the problems could be solved as written and talked to each other about what had been written. During this time I walked around the room listening in on some of the conferences between students. Their comments ranged from "Don't you need a question mark at the end of this sentence?" to "How many Cokes did you order?" and "I don't think you can solve this problem!" The student pairs were engaged in both proofreading and problem-solving activities. It was easier for them to detect errors when they were looking at someone else's work. They were more convinced of the lack of clarity in their own writing when they discussed their writing with their peers than when the teacher pointed out mistakes and asked them to make changes.

Discussing what was wrong with word problems helped the students to look at the structure of the problem, examining details and focusing on the question being asked. Little intervention by the teacher was needed during the conference. The students were much more thorough in their examination of each other's writing than I expected. When the students were satisfied that they had an acceptable problem, they returned to their own seats to move into the next stage of the writing process: revising and editing.

Revising and editing

The comments received from their peers during the conference stage made revising and editing an easy task. Students made the corrections on their own problems by crossing out what they did not want to keep, drawing arrows to rearrange what they had written, and even cutting the sentences apart and taping them back together in the order that they decided on during the conference with their peers. They were then ready to put their word problems into final form.

Publication

The motivation to go through the conference, revising, and editing stages came from the last stage of the writing process: publication. The students in this particular classroom were setting up a file box of problems to be located on the mathematics table. They knew that the other students would be reading and trying to solve their problems. They each took a 4 x 6 index card and copied their problem neatly onto the card and recorded the solution to the problem on the back. The students then signed their names to the bottom to indicate authorship of the problems they wrote. If a student had difficulty in solving one of the problems in the file box, he or she could contact the author to discuss how to proceed to a solution.

Follow-up

Some students immediately began taking cards out of the file box to attempt solving the problems that had been written by their friends; still others asked to write more problems. One of the students asked if he could write his own word-problem book, which he intended to call "David's Problems." Both the teacher and I were amazed to see students excitedly engaged in problem-solving activities. The process did not end with this work session. The teacher reported that the students were choosing to work on the problems written by their peers long after their introduction to the process, writing new ones, and adding to the collection of problems in the file box. Solving the problems in the box was proudly announced at various times throughout the school week. Here are some examples of the students' work:

> One day my mom wanted to go to the store to get something for work. She took me along. I had $100.00 and I wanted to buy a portable tape player for $30.00, a portable radio for $64.00, and a clock radio for $29.00. Did I have enough money? (Bryan)

> One day my dad took me to a Chinese restaurant. He ordered Barbecue Ribs. They cost $4.98. He only had $2.75. How much does he need? (Brad)

> Two food processors cost $73.88. How much does only one cost? (Becky)

> My dad said one afternoon, Let's go to the TV shop. When we got there, he found out he only had $79.00. The TV cost $499. How much more money does he need? (Krisdee)

The problems written by these third graders were not problems that they had typically been asked to solve. Some were multistep problems involving addition and subtraction, and some involved multiplication or division. The students did not model problems that they had seen in textbooks. They were actively engaged in the process and in recording their own

thinking. Their problems reflected real situations. The students applied their understanding of the mathematics they had learned to solve problems that were unique to them and their friends. Problem solving was becoming meaningful to them.

After visiting the third-grade classroom and sharing my findings with other teachers, a sixth-grade remedial-mathematics teacher decided to try the writing process in her classroom. Many of her students were unsuccessful in mathematics and were older than the typical sixth grader. The teacher was frustrated by their inability to perform simple computation and by the attitude displayed by the students toward mathematics and school in general. After writing word problems using the writing process, these sixth graders, like the third graders, were eager to solve the problems that their friends had written. They wrote problems that they could not (or would not) have solved had they been assigned from a textbook. The teacher was surprised to see these students eagerly writing and competing with each other to solve their friends' problems. This class decided to write most of their problems for another sixth grade class in the building, so two classrooms were writing and exchanging problems. Some of their problems follow:

> Jack went fishing on Mondays, Wednesdays, and Fridays for 6 weeks. How many days did he go fishing? (Tameka)

> Michelle bought a pair of pants for $14.99. She also bought a jacket for $24.99. She had $50.00. Would she have enough money to buy a sweater for $8.99? (Connie)

> Devin went to Roses and bought 3 rugs for $10.44 each, an ironing board for $9.98, and shot gun shells for $3.67. What was his change from $60? (Clinton)

Writing problems and solving them gives students the experience they need to develop and refine their problem-solving skills. Students must be given the opportunity to write their own word problems using the writing process on a regular basis and must be allowed time to work on problems that others have written. Using the writing process as a strategy for problem solvers seems to help students focus on the question being asked, look for essential information, and become familiar with the structure of the written problem. The eagerness of students to engage in problem-solving activities is both exciting and rewarding! If students willingly attack problems written by their peers and are successful in solving those problems, perhaps textbook story problems can be replaced by those written by students.

The writing process is a successful strategy for problem solvers if students move through all the stages of the writing process and are furnished with a calculator for solving problems. If we are to make progress toward the goal of developing the problem-solving abilities of our students, then the writing process can be only one strategy that teachers employ in teaching problem solving. It deals primarily with solving routine story problems and cannot replace instruction in problem-solving strategies and experiences with all kinds of word problems. Without a doubt, problem solving needs attention in the elementary school. Through writing, let us begin by helping change students' attitudes and by giving them a sense of ownership over their own learning in problem solving!

Bibliography

Burton, Grace M. "Writing as a Way of Knowing in a Mathematics Education Class." *Arithmetic Teacher* 33 (December 1985):40-45.

Calkins, Lucy. *The Art of Teaching Writing*. Portsmouth. N.H: Heinemann Educational Books, 1986.

Davison, David M., and Daniel L. Pearce. "Using Writing Activities to Reinforce Mathematics Instruction." *Arithmetic Teacher* 35 (April 1988):44-45.

Fennell, Francis (Skip), and Richard Ammon. Writing Techniques for Problem Solvers. *Arithmetic Teacher* 33 (September 1985):24-25.

Ford, Margaret I. "Fifth Grade Teachers and Their Students: An Analysis of Beliefs about Mathematical Problem Solving." Ph.D. diss. University of South Carolina. 1988.

Geeslin, William E. "Using Writing About Mathematics as a Teaching Technique." *Mathematics Teacher* 70 (February 1977): 112-15

Graves, Donald H. *Writing: Teachers and Children at Work*. Exeter, N.H.: Heinemann Educational Books. 1983.

Johnson, Marvin L. "Writing in Mathematics Classes: A Valuable Tool for Learning." *Mathematics Teacher* 76 (February 1983): 117-19.

Nahrgang, Cynthia L., and Bruce T. Petersen. "Using Writing to Learn Mathematics." *Mathematics Teacher* 79 (September 1986): 461-65.

National Council of Teachers of Mathematics, Commission on Standards for School Mathematics. *Curriculum and Evalualion Standards for School Mathmatics*. Reston. Va.: The Council, 1989.

Suydam, Marilyn N., "Untangling Clues from Research on Problem Solving". In *Problem Solving in School Mathematics,* 1980 Yearbook of the National Council of Teachers of Mathematics, edited by Stephen Krulik and Robert E. Reys, 34-50. Reston, Va.: The Council, 1980.

Watson, Margaret. "Writing Has a Place in a Mathematics Class." *Mathematics Teacher* 73 (October 1980):518-19.

Students Can Write Their Own Problems

Patricia C. Hosmer

For years, my class of third graders has written their own holiday problems, which I have typed and copied so all classmates could enjoy their friends' work. We had had the standard problems about Santa and Christmas trees, menorahs and dreidls, so this past year I decided to vary the format. I asked the children to do a little research on holiday customs in one of the countries of their ancestors and then to develop a mathematically related question about the custom. For many students the assignment involved a trip to the library. After it was completed, several parents commented, "I didn't know anything about holiday customs in Italy, Russia, ..." Then the problems came in; some were quite long and involved, but in almost all the child's own words were clear and needed little or no revising. Some children asked one question; several asked three!

Most children wrote a paragraph explaining the custom before they presented the problem. My class was a microcosm of America, so I learned about Diwali, the Indian festival of lights; Christmas customs in China, Ethiopia, central Africa, Italy, Ireland, Germany, and the Ukraine; and Hanukkah customs in Poland and Russia.

Some problems were complicated, and not all children were able to solve the most difficult ones. The problems were written so that their solution required all the mathematical operations we had studied, through simple division. The most difficult problems demanded careful reading and required a solution of several steps.

Some planning is needed if this activity is to succeed. During the fall the third-grade students had

been writing "thought problems" for each other. I had written some paragraphs to go with our history lessons, and I asked children to write four problems using data in the paragraphs. These problems also showed a wide range of mathematical understanding. This experience encouraged me to go further and ask children to write a mathematical problem based on a book that we were reading in class or one they were reading for pleasure at home or based on some numbers pertinent to their family life. Children used their addresses, phone numbers, and ages: one used his dog's license number in a problem.

Almost daily, mathematics instruction in my class involves problems that I dictate to the students. We discuss the solutions, noting the various methods of solving them This discussion gives me insight into how children work problems, and I often learn why they have certain difficulties by watching them work in small groups and by listening to them explain their results.

Young children must have a wide variety of experiences in writing before they can develop the skills needed for clear, concise explanations. The teachers at our school put great stress on problems, first presenting them orally and then in writing, finally having children construct their own problems to share with others.

The results are being seen in other disciplines, and we expect some transfer to instruction in computer languages.

Now for a few selected problems by some of my third graders:

1. In Germany, Advent lasts for twenty-four days. In the von Trapp family each sister has the same number of sisters as brothers and each brother has twice as many sisters as brothers. Each child gets one present each day of Advent. How many presents do all the children get? (Carwil James)

2. Hanukkah lasts for eight days and nights. On the first night you use two candles (Hanukkah candles burn so fast you have to use new candles every night.) On the second night you add a candle. Each night you add one more candle. The custom is to light two menorahs every night if you have two children. Hanukkah candles come fifty in a pack. If you have two children, how many more candles will you need than are in one package of candles? (Becky Davis)

3. If two pounds of potatoes make ten latkes (potato pancakes), how many pounds will we need to make thirty latkes? (Erica Lieberman)

4. In China the Christmas festival is known as Sheng Dan Jieh, the Holy Birth Festival. Many kinds of paper decorations and evergreens fill the churches and homes. The Chinese call the Christmas tree the tree of light. They don't use candles, but the tree is decorated with three dozen paper flowers, five dozen colored paper chains, and two dozen cotton snowflakes. How many ornaments are placed on the tree all together? (Chris Davis)

5. During the Christmas season in Italy, families that have mangers set them up. Musicians sing before the manger, and guests kneel before it. In one town, there were thirty-seven houses and twenty-one had mangers. In each house that had a manger, two musicians sang to it. Five guests also knelt in front of it. (a) How many guests and musicians are in this town? (b) How many more houses still need mangers? (c) How many guests and musicians would be visiting this town if every family had a manger? (Patrick Quigley)

6. For Diwali (an Indian festival of lights) I bought forty fireworks. Each was supposed to give five sparks. and then explode once. Unfortunately, three did not work at all, five only sparked, and four only exploded. (a) How many sparks appeared? (b) How many explosions occurred? (Arun Shivashankaran)

7. The Irish celebrate St. Stephen's Day on 26 December. A group of boys sang eight Christmas carols at each house. Each carol had five verses. The boys sang at ten houses. How many verses did they sing? (Colleen Shanahan)

8. Ukrainians celebrate Christmas on 6 January. For thirty-nine days before the holiday, people observe a partial fast. On what day do they start their fast? (Sarah Janicki)

CHAPTER 5

Programming and Problem Solving in Logo

In the early days of personal computing, being computer literate meant knowing how to program. And, for a time, it seemed that anyone with good programming skills would always be able to find a job as a computer programmer. Today, the need for computer programmers has leveled off, since most business and personal computer needs are met with general-purpose computer tools such as word processors, spreadsheets, and data base managers rather than with programs written to satisfy a specific need. Indeed, many are questioning whether anyone except college students who plan to become computer specialists should learn to program.

The argument that only prospective computer professionals should learn programming is similar to insisting that only those who plan to become professional basketball players should play basketball in school physical education classes—why not concentrate on aerobic exercises and "lifelong" sports such as golf, hiking, and cycling? Of course, we do not teach and play basketball in physical education courses for the purpose of developing future professional basketball players. Rather, we teach it because it is a highly enjoyable way to exercise the cardiovascular system, it develops teamwork, it improves children's hand-eye coordination, and so on.

Likewise, the experience of programming a computer has many benefits for the learner without considering any career possibilities. One benefit is that, when programming, you are learning how computers work at an up-close level. When working with applications programs, especially in user-friendly environments such as most Macintosh applications, students are far removed from the actual workings of the machine. When programming, however, students are learning general concepts that are fundamental to all computer applications. They are also applying a variety of mathematical concepts—variables, numeration systems, order of operations, and coordinate graphing systems, to name a few.

Another benefit is that the act of programming is almost synonymous with problem solving. Unless you are simply copying or making minor changes in someone else's program, when developing a computer program to do a particular task you must (1) understand the problem, (2) develop a plan for solving it (plan the program), (3) carry out the plan (write the program), and (4) check your answer (test and debug the program). Educators such as Seymour Papert (1980) strongly feel that the strategies learned and used when writing computer programs are generalized and transferred—they have an impact on the way we solve other problems. For example, if you have analyzed and understood computer problems and have planned computer programs, you are more likely to analyze and understand and carefully plan your solution for noncomputer problems.

Computer problem solving—programming—is more specific than general mathematical problem solving. First of all, not all mathematical problems lend themselves to computer analysis. For example, many logic problems (e.g., "John is Sally's sister, Susan plays squash with Judy, John once dated Judy, . . ."), spatial problems (e.g., "Arrange 11 matches to form 3 squares"), and problems involving numbers (e.g., "I have $1245.73 and the new computer I want to buy costs $2145.09—How much more money do I need?") are best solved by other means. And, even though Logo's Turtle Geometry was developed specifically for solving problems in geometry, only a small subset of geometry problems are solvable via Logo. Indeed, many of the problems one does in Logo are not even problems outside of Logo (e.g., in traditional geometry, we are more concerned with *defining* squares and looking for "examples" of squares, than with drawing them).

However, there is a wealth of mathematical problems, both geometric and nongeometric, where pencil and paper or calculators are not appropriate and where spreadsheets or data base managers do not provide sufficient flexibility. Programming languages can be used to tap any of a computer's capabilities. And languages such as Logo and BASIC are simple enough to learn and use that they are within the range of elementary children—Logo is used even by preschool and kindergarten children to solve problems.

Whether in BASIC, Logo, or some other programming language, computer-based programming/problem-solving activities add a great deal of excitement to the classroom, reinforce many mathematical concepts, develop an understanding of computers, and provide a powerful tool that can be used in a variety of problem situations. If problems are

---

Programming/Problem-Solving Heuristics

1. Understand the problem
   Restate it in your own words
   Write the problem in your own words
2. Plan your solution
   Describe (in English) the major parts of your proposed
      computer program
   Write your program in pseudocode (English phrases that
      describe the various parts of the program)
3. Carry out the plan
   Write the main module of the program as a simple
      sequence of calls to the modules of the program
      (subroutines or procedures where the actual detailed
      program code will occur)
   Write and test the subroutines or procedures in the program
4. Test and debug the program

Fig 5.1: Programming/problem-solving heuristics.

properly chosen and problem solving heuristics are emphasized (see Fig. 5.1) children will be using problem-solving strategies that are applicable in other, noncomputer areas as well.

There are many programming languages available for microcomputers. Besides Logo and BASIC, the most commonly taught programming language at the pre-college level is Pascal, primarily to upper-level high school students in advanced placement (AP) computer science courses. Pascal was developed specifically for teaching structured programming concepts to computer science students. There are many Pascal packages—such as Turbo Pascal and Quick Pascal—that have excellent editors and other features (e.g., graphics tools, tutorial disks). But, even though there are simplified versions of Pascal (e.g., Apple's Instant Pascal) and packages designed to gently introduce Pascal concepts to younger students (e.g., *Karel the Robot* [Pattis, 1981]), Pascal is rarely taught below the high school level or in introductory computer courses.

Two other languages should be mentioned: Cobol is occasionally taught in pre-college business classes. The programming language C (a powerful language based on Pascal) is occasionally taught in advanced high school computer science courses. But of all the languages available, Logo and BASIC are, by far, the most commonly used and the best suited for the kinds of explorations you might wish to have students do in the mathematics classroom. The remainder of this chapter introduces Logo and Logo programming activities for the elementary (K-8) mathematics classroom and the fundamentals of structured programming. Chapter 6 introduces BASIC and its applications in the elementary, middle-grade and high school mathematics classroom.

Solving Mathematical Problems in Logo

The programming language Logo was developed specifically to provide children with a tool for exploring mathematics and other topics. Logo contains a built-in vocabulary that allows the programmer to do a variety of mathematical tasks. Logo also allows the user to do a variety of nonmathematical tasks. For example, the programmer can write a Logo command that asks the computer to solve a mathematical expression and print the result of the computation. Logo provides the means to store values and use them in later computations, to repeat actions as often as desired, to make and to manipulate lists, to define new commands, and to draw shapes on the screen. All of these capabilities can be useful in exploring mathematics. Paul Ernest, in his article at the end of this chapter, gives an overview of the uses of Logo in the mathematics curriculum. He also discusses the instructional philosophies and methodologies related to its use. In her article, Patricia Campbell presents a rationale and a set of strategies for using Logo with kindergarten, first- and second- grade children. She discusses single-keystroke versions of Logo designed for use by young children. Diane McGrath, in "Eight Ways to Get Beginners Involved in Programming." gives further suggestions about teaching programming to learners of any age—from kindegarten to adult. She focuses on classroom strategies that are motivational. Dorothy Keane, in her article, gives suggestions for using Logo to enhance children's spatial problem-solving capabilities.

Perhaps the best known and most frequently used component of Logo in the mathematics classroom is its graphics language, Turtle graphics, a child-oriented tool for exploring geometry. According to Seymour Papert (1980), its co-developer and chief spokesman,

> [when children control the Turtle using Logo, they] are learning a language for talking about shapes and flux of shapes, about velocities and rates of change, about process and procedures. They are learning to speak mathematics, and acquiring a new image of themselves as mathematicians (p. 13).

In this section we first look at Logo commands and procedures that allow us to program the computer to do mathematical computations. Next, we look at Turtle graphics and their use in the mathematics classroom.

Logo Commands

The basic building blocks of Logo are *primitives, statements,* and *procedures* [1] (see below). A primitive is a built-in command to the computer to carry out a specific task (e.g., draw a line on the screen). One of the most useful and easiest to learn primitives is "PRINT." This command tells the computer to print something on the screen. If a pair of brackets follows the PRINT primitive, the computer prints the sequence of characters between them on the screen. For example,

```
PRINT [Logo is a friendly language]
```

prints the phrase "Logo is a friendly language," on the screen.[2]

PRINT can also be used to ask the computer to do mathematical operations and print the result. For example,

```
PRINT 4 + 5 * 7
```

carries out the operation "4 + 5 * 7" then prints the result, "39", on the screen. Note that in Logo, as in most computer languages, the multiplication operation is symbolized by an asterisk (*).

When computing the result of mathematical expressions, the computer follows the standard order of mathematical operations; namely, it does all multiplications and divisions first from left to right, then it does all additions and subtractions from left to right. Thus, in the expression "4 + 5 * 7," the

---

[1] A statement is like a sentence. It contains a primitive or command (such as PRINT). Depending on the type of command, it may also contain a value, expression or the like, which the command acts on (e.g., PRINT 3*5). In most versions of Logo, statements are terminated by pressing the RETURN (or ENTER) key and may not be broken up by carriage returns.

[2] There are several versions of Logo in the market. The Logo statements printed in this book will run on the Terrapin Logo (also called Logo Plus) version of the language. Most of the primitives and statements presented in this chapter will run on any version of Logo. If a particular statement does not run with the verison of Logo you are using, check your reference manual.

computer first carries out the "5 * 7" operation, then the "4 + 35" operation. Figure 5.2 gives a more complex example.

Parentheses may be used to override this standard order of operations. The computer does operations in parentheses first (from left to right). If parentheses are embedded inside of parentheses, it does the innermost first (Figusre 5.3).

When you first load Logo into the computer's memory, most versions of the language will print a ? on the screen and wait for you to type something. If you type a PRINT statement such as one of those above and press RETURN, the computer will execute it immediately, print the results on the screen, then wait for you to enter another command. Fig. 5.4 shows what the screen looks like (in most versions of Logo) when *immediate commands* are executed.

Logo Procedures

Children's initial experiences with Logo, especially when working with Turtle graphics, will be in the immediate mode. The limitations of the immediate mode, however, will quickly become apparent, if not to your pupils, at least to you: after you have executed a command, there is no way to retrieve it or edit it. For example, if while creating a complex shape on the screen you discover that you need to change one or more of the previously executed commands in order for your shape to be correct, you are out of luck. Or, if you wish to create a set of commands to use several different times without retyping them each time, again you are out of luck. Fortunately, Logo (and all other programming languages) allows you to enter and edit sets of commands that are executed as a unit and not until the user commands the computer to do so. A set of Logo statements that is intended to be executed together is called a *procedure*. Procedures begin with TO, followed by the procedure's name, and end with END.

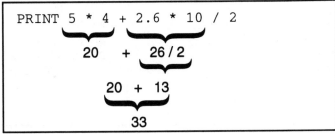

Fig.5.2: Order of operations in Logo.

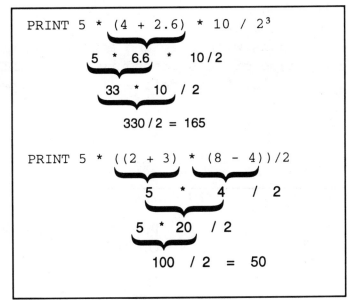

Fig. 5.3: Order of operations with parentheses

```
?PRINT 5*(4+2.6)*10/2
165
?PRINT 5*((2+3)*(8-4))/2
50
?PRINT [HELLO, HOW ARE YOU?]
HELLO, HOW ARE YOU?
```

Fig. 5.4: Immediate commands in Logo

---

[3] In Logo, as in other programming languages, the operation sign between a number and a left parenthesis sign must be explicitly written. Thus, 3 times the sum of 4 and 5 is written "3*(4+5)," not 3(4+5) as is is in algebra.

```
TO TEACHME
    PRINT [This is a procedure.]
    PRINT [It begins with a TO and ends with an END.]
    PRINT [The name of this procedure is TEACHME.]
    PRINT [To run it, first define it, then type TEACHME.]
END
```

```
?TEACHME
This is a procedure.
It begins with a To and ends with an END.
The name of this procedure is TEACHME.
To run it, first define it, then type
TEACHME.
?
```

Fig. 5.5: Output from PRINT procedure

Most versions of Logo put you into a procedure editor as soon as you type TO followed by a name and press the RETURN button. As soon as you type END and press RETURN, your procedure is defined. To run a procedure, simply type its name and press RETURN. Fig. 5.5 shows the output when the procedure above is executed.

Consult your reference manuals to learn how to define and edit procedures in your version of Logo.

Pupil Activity 5.1: Printing Text and Operations in Procedures
Objective
Write procedures to print text and do mathematical operations.
Grouping
Individuals or pairs of pupils.
Materials
Computer, Logo.

Procedure
1. Ask students to write a short poem or humorous saying (4 to 10 lines long) using PRINT statements in a procedure. Have them run it and, if necessary, edit it so that it runs correctly.
2. After discussing order of operations in arithmetic, give students a set of problems to do involving several operations. Have them calculate the answers using mental arithmetic, pencil and paper, or a calculator. Next, have them use PRINT statements in a procedure to check their answers.
   Problem examples:
   a. 6 + 2 + 9 - 5 * 8
   b. 12 / 2 - 4 * 10 * 10
   c. (6 + 2) * 4 - (2 + 13) * 2

Mathematical Operations in Logo

In addition to the mathematical operations discussed above, Logo contains a set of primitives for doing operations in arithmetic. For example, if you wanted to find and print the square root of a number, you would type the following statement:

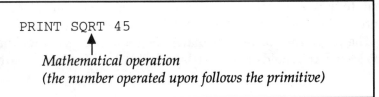

```
PRINT SQRT 45
```
Mathematical operation
(the number operated upon follows the primitive)

Fig. 5.6: Mathematical primitive

Like any other operation (e.g., "3 + 4"), a mathematical primitive is not a complete statement. It must be used as a part of a complete Logo statement such as the following:

```
PRINT 3 + 4 * SQRT 6
```

Be careful if mathematical operations *follow* the mathematical primitive in an expression. Depending upon which version of Logo you use, the expressions after the primitive may be evaluated before the operation is executed. Thus,

```
PRINT SQRT 1 + 1 + 1
```

in many versions of Logo will add the three 1s, then find and print the square root of 3. To avoid this problem, enclose the operation in parentheses:

```
PRINT (SQRT 121) + 1 + 1
```

will find the square root of 121, then add the two 1s to the result.

Some of the other mathematical operations executed by Logo are as follows:

INTEGER 3.5678	calculates the integer part of the number—3.
RANDOM 20	calculates a random number from 0 to 19, or one less than the number given.
REMAINDER 19 5	calculates the remainder if the first number is divided by the second—4 in this example.
ROUND 4.667	rounds to the nearest integer—5 in this example.

Variables in Logo

Another important primitive, MAKE, allows you to store values in variables for later use. The following statement assigns the value of 5 to the variable VAR1:

```
MAKE "VAR1 5
```

If the computer had previously executed this command, the contents of VAR1 could be printed as follows:

```
PRINT :VAR1
```

The colon signals to the computer that the *value* stored in the variable VAR1 should be printed. The statement PRINT "VAR1 will print the variable's name, VAR1. The procedure below calculates the total price paid for the food in a Happy Meal Party, held on the last day of school in Mrs. Jones' sixth grade class.

```
TO PARTY
   MAKE "COST 2.95
   MAKE "NUMBER 33
   MAKE "TAX .065
   MAKE "PRICE :COST * :NUMBER + :COST * :NUMBER * :TAX
   PRINT (SE [The cost of 33 Happy Meals is $] :PRICE)
END
```

Note the format of the PRINT statement: if you wish to print more than one entity (the set of characters between the brackets is one string, a variable is another entity, an expression [e.g., "3 + 4 * 5"] is another, etc.) after PRINT, you must combine them in a *sentence*. A sentence is enclosed in parentheses and begins with SE.

If you wish to add to or otherwise modify the value stored in a variable, you may combine in one statement the steps of modifying the value of the variable and replacing the old value of the variable with the new. The following line adds 5 to the value in VAR5, then replaces the old value with the new:

```
MAKE "VAR5 :VAR5 + 5
```

If, previously in the procedure, the value of 100 had been assigned to VAR5, this line would add 5 to 100 to get 105, then store 105 in VAR5. To see this new value, we can use the following statement:

```
PRINT :VAR5
```

Putting it all together into a procedure:

```
TO EXAMPLE2
   MAKE "VAR5 100
   MAKE "VAR5 :VAR5 + 5
   PRINT :VAR5
END
```

Pupil Activity 5.2: Variables in Logo

Objective
Use variables in procedures.
Grouping
Individuals or pairs of pupils.
Materials
Computer, Logo.
Procedure
Ask the students to write procedures using variables to calculate and print the results of problems like the following. Of course, problem-solving

strategies should be used (e.g., understand the problem, plan your solution, and so on):

1. Al earns $2.40 an hour babysitting. Last month he babysat for a total of 43 hours. How much did he earn?

2. The seventh-grade class at Belmont School collected 6328 pounds of newspaper in their recycling fund raiser for their class trip. If City Recyclers will pay them $75.95 a ton for newspapers, how much will they earn?

REPEAT

The REPEAT primitive allows you to direct the computer to repeat a statement or set of statements a given number of times. The format of the REPEAT statement is as follows:

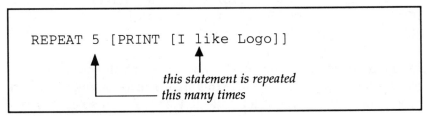

Fig. 5.7: The REPEAT command

This program line will print "I like Logo" 5 times on the screen. Using variables in REPEAT statements provides a powerful way to produce number patterns and to do a variety of computations. The following program prints the numbers from 5 to 200 by 5s:

```
TO FIVES
    MAKE "VAR1 5
    REPEAT 40 [PRINT :VAR1 MAKE "VAR1 :VAR1 + 5]⁴
END
```

Suppose you wanted to know how much money you would collect after 25 years if you invested $5000 in a savings bond that earned 8% a year until its maturity 25 years later. This program will calculate the answer:

```
TO EARN
    MAKE "MONEY 5000
    REPEAT 25 [MAKE "MONEY :MONEY + :MONEY * .08]
    PRINT (SE [After 25 years, the bond is worth $]:MONEY)
END
```

Decisions

Sometimes, we do not know how many times we want a loop to be executed, or we want to carry out some action based on the value of a variable or the value entered from the keyboard. For this type of task, we use the IF

---

⁴ The REPEAT statement carries out two different tasks, printing (PRINT :VAR1) and assigning a value to VAR1 (MAKE "VAR1 :VAR1 + 5)

command. The IF evaluates a conditional statement (one that may be determined to be true or false) and carries out an action based on the result. The form of the IF statement is as follows:

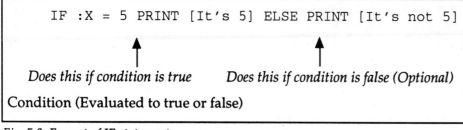

Fig. 5.8: Format of IF statement

If the condition :X = 5 is true, the statement or set of statements after the condition is executed. Otherwise, the statement(s) after the ELSE is (are) executed. Some examples of conditionals:

```
IF :VAR1 < 20      (The value of :VAR1 is less than 20)
IF :COUNT3 > 40    (:COUNT3 is greater than 40)
IF NOT (:M = :N)   (:M is not equal to :N)
```

More complex conditionals, using ANDs and ORs, are also possible. See your reference manual for further help in this area. The following procedure checks to see if enough cans have been collected (at 2 1/2 cents a can) to earn the amount needed ($220) to admit a class to Swift Waters Amusement Park for their end-of-year Fun Day.

```
TO CHECK
    MAKE "NUMCANS 5248
    MAKE "MONPERCAN .025
    MAKE "MONEY :NUMCANS* :MONPERCAN
    IF :MONEY > 220 PRINT [You have collected enough cans]
    ELSE PRINT [You need to collect some more cans]
END
```

The IF statement can also be used to check if a condition has been met in a loop. If, for example, you wanted to print the number sequence, 7, 14, 21, 28, . . . up to 100 (but did not wish to do the arithmetic to determine the number to place after the REPEAT command), put a sufficiently large number after the REPEAT, then use the IF to stop the looping.

```
TO NUMS
    MAKE "X 7
    REPEAT 100 [PRINT :X  MAKE "X :X + 7 IF :X > 100 STOP]
    PRINT [End of loop]
END
```

The REPEAT will keep looping in this procedure until the value of :X is equal to or greater than 100, then it will stop looping and the next statement after the REPEAT will be executed.

Pupil Activity 5.3 Solving Problem with IF and REPEAT

Objective
Use IF and REPEAT to solve problems.
Grouping
Individuals or pairs of pupils.
Materials
Computer, Logo.
Procedure
Solve the following problems using variables, IF, and/or REPEAT in procedures:

1. Write a procedure to print the numbers from 100 to 10,000 by 100s. (Use IF to stop the procedure after it has printed 10,000.)
2. Write a procedure to print the numbers from 280 down to 120 by 4's. (Use IF to stop the loop.)
3. Modify the can collection program above so that, if not enough cans have been collected to be worth $100, it tells you how many more cans are needed.
4. Wilson school is converting a classroom into a library. The school bought 20 wooden bookcases for the new library. Each bookcase has 5 shelves and each shelf can hold 30 average-sized books. The school has 3628 books and wants to have enough space for 1000 more books. Write a procedure to find out (and print the answer to) whether the school has enough bookcases. If not, the procedure should print out how many more bookcases it needs to buy.
5. John got a job taking care of the Smith's yard and plants for twelve weeks while they were staying at their cabin in Minnesota. Mrs. Smith gave John a choice of two types of payment: (1) $5 a week for the 12 weeks, or (2) 10 cents the first week, 20 cents the second week, 40 cents the third week, 80 cents the fourth, and so on. Write a computer program to find out how much he would earn if he chose the second option. Which is the best choice? How much would John earn using this method of payment if the Smiths were gone for a year (52 weeks)?

Getting Input from the Keyboard

Suppose you needed to convert a large number of distances expressed as kilometers into miles. Or, suppose you wanted to find the total cost of a set of gifts purchased through a mail-order retail company and sent to friends and relatives in another part of the country (cost of the gifts, taxes, wrapping, and shipping). Of course, Logo would be an appropriate tool for these tasks. And, in either case, you would want to write a general program that would work with any fraction or gift by allowing you to input the appropriate values from the keyboard. Fortunately, Logo has a primitive, REQUEST (READLIST in some versions of Logo) that halts a procedure and waits for the user to type a set of characters from the keyboard. This input may then be stored in a variable for later use. The following statement assigns the characters typed from the keyboard to the variable "NUM.

```
              MAKE "NUM FIRST REQUEST 5
   or         Make "NUM FIRST RQ
```

The following procedure converts a length in kilometers inputted from the keyboard into equivalent distance in miles.

```
TO FRACTION
   PRINT [Please type a distance in kilometers]
   MAKE "DIST FIRST RQ
   PRINT (SE :DIST [ kilometers equals ] :DIST /1.609
          [ miles])
END
```

Suppose that you wanted to run this procedure an unknown number of times—long enough to convert a large set of distances in kilometers to miles. You might want to call up this procedure within another procedure until the user types something indicating he/she wants to stop (once defined, a procedure's name may be used as a new Logo command).

```
TO CALC
   REPEAT 1000 [FRACTION PRINT [To stop type "S", to continue
   press any other key] MAKE "INP READCHARACTER 6 IF :INP =
   "S" STOP]
END
```

Procedures with Inputs

On previous pages in this chapter, we presented the MAKE primitive and its use in assigning values to variables. Another way to assign values to variables is to create procedures that accept inputs. For example, if you wanted to create a procedure to multiply any two numbers together, you could add two variable names after the name of the procedure, then use the variables in the body of the procedure:

```
TO MULT :NUM1 :NUM2
   MAKE "NUM3 :NUM1 * :NUM2
   PRINT :NUM3
END
```

To execute this procedure, type its name followed by two values:

```
MULT 5 8
```

The computer will execute the procedure and output the results (40). And, as with any other procedure, you may use it alone or as a part of another procedure:

---

[5] The primitive FIRST must be typed before REQUEST in the situations presented here. The reason for this will not be explained in this text. Consult your reference manual if you desire further explanation.

[6] The primitive READCHARACTER gets a single character from the keyboard.

```
TO CALCULATE
   PRINT [This procedure allows you to multiply two
         numbers]
   PRINT [Please type the first number]
   MAKE "FIRSTNO FIRST RQ
   MAKE "SECONDNO FIRST RQ
   MULT :FIRSTNO :SECONDNO
   PRINT [That's all]
END
```

Pupil Activity 5.4

Objective

Use keyboard input in procedures.

Grouping

Individuals or pairs of students.

Materials

Computer, Logo.

Procedure

Solve problems similar to the following:

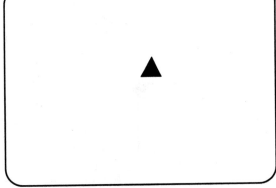

Fig. 5.9: Turtle in home position

1. Write a procedure which gets the height, base, and type of unit of a triangle, then prints out its area followed by the word "square" and the type of unit.
2. Write a procedure which gets two numbers from the keyboard—the first is any integer or rational number, the second is 10, 100, 1000, 10,000, etc. (a power of 10). The procedure then prints the product of these two numbers.[7]

Turtle Graphics

Not only does Logo allow the user to direct the computer to do mathematical computations, print, assign values to variables, and do repeated tasks, it also contains a powerful graphics language called Turtle graphics. This graphics language was developed with the young learner in mind. Rather than using an absolute, abstract coordinate system for positioning shapes on the screen, it uses a more concrete, learner-oriented system. The graphics screen contains a small triangle—a "turtle"—that the child controls by typing commands from the keyboard. All movements of the turtle are relative to its current position and heading on the screen. Thus, children can move around "screens" marked on the floor or control robot toys to help understand the movement of the turtle on the screen.

The turtle can be programmed to move forward or backward a given number of "turtle steps," or to spin to the right (clockwise) or left (counterclockwise) a given number of degrees. When entering the graphics mode (some versions of Logo enter the graphics mode as soon as the computer executes a graphics command whereas others have a specific command that causes the computer to leave the text mode and enter the graphics mode), the turtle appears in its "home position"—the center of the screen, pointing up (Fig. 5.9).

---

[7] This procedure could be used by students to experiment with the effect of multiplying a number by powers of 10 and, hence, to discover an important mathematical principle/ shortcut (moving the decimal point to the right in a number). A parallel activity would be to write and use a procedure that *divided* an inputted number by powers of 10.

The first graphics statement encountered by the computer after entering the graphics mode, then, causes the turtle to move, as directed, from this beginning position. If, for example, you entered the following statements:

FORWARD 90   (this can be shortened to FD 90)
LEFT 45   (this can be shortened to LT 45)

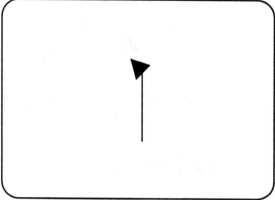

the turtle would move straight ahead 90 steps (pixels[8]), then turn to the left (counterclockwise) 45 degrees. (Fig. 5.10).

If, subsequently, you entered the following commands:

BACK 60   (this can be shortened to BK 60)
RIGHT 90   (this can be shortened to RT 90)

the turtle would move backward 60 steps, then spin to the right 90 degrees. (Fig. 5.11).

Fig. 5.10: Logo screen with FORWARD and LEFT

The following procedure would draw a star on the screen.

```
TO STAR
     REPEAT 5 [BACKWARD 75 RIGHT 145]
END
```

After you have drawn on the screen using immediate commands or procedures, you may wish to clear the screen, start over, or move the turtle back to its home position. The following commands take care of these housekeeping tasks for you. They may either be used in the immediate mode or within procedures.

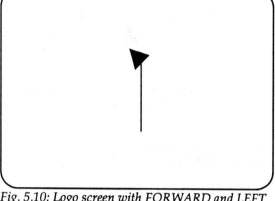

HOME   (Sends the turtle back to its home position)
CLEARSCREEN or CS (Clears the screen and sends the
                   turtle back to its home position).

Fig. 5.11: Logo Screen 3

If you would like to draw separate shapes on the screen, the PENUP (or PU) command will tell the turtle to retract its "drawing pen" so that it can move around the screen without leaving a line on the screen as it moves. Once the turtle is in a position to start drawing again, PENDOWN (or PD) will cause the turtle to start drawing again. Thus, the following procedure would draw two squares on the screen as shown in Fig. 5.12.

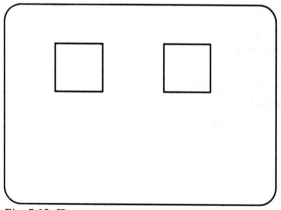

Fig. 5.12: Two squares

---

[8] A pixel is a programmable point on the screen. The number of pixels on a screen varies according to the type of graphics display used by a particular computer.

```
TO SQUARES
   PU
   FD 50
   LT 90
   FD 50
   PD
   REPEAT 4 [FD 70 RT 90]
   PU
   BK 100
   RT 180
   PD
   REPEAT 4 [FD 70 LT 90]
END
```

Recursion

A powerful feature of Logo is the ability of a procedure to call and execute itself. For example, you might wish to create a series of squares, each slightly larger than the previous one and rotated slightly. To accomplish this task, you could create a procedure with variable inputs that calls itself (Fig. 5.13).

```
TO REPSQUARES :SIZE :ADD :BEGCT :FINCT
   MAKE "BEGCT :BEGCT + 1
   IF :BEGCT = :FINCT TOPLEVEL⁹
   REPEAT 4 [FD :SIZE RT 90]
   MAKE "SIZE :SIZE + :ADD
   RT 10
   REPSQUARES :SIZE :ADD :BEGCT :FINCT
END
```

Fig. 5.13: Recursive procedure

To use this procedure, you type its name followed by the original size (side length) of the square, the amount to be added to the length of the sides each time the procedure is executed, the beginning count, and the ending count. Thus,

```
REPSQUARES 5 8 0 10
```

would draw a 10 squares of side length 5, 13, 21, and so on, each rotated 10 degrees clockwise from the previous square (See Fig. 5.14). Notice the third line of this procedure: When the count (BEGCT) is equal to the final count (FINCT), the TOPLEVEL primitive is used to stop all procedures. [9]

The primitives and techniques just introduced are but a small fraction of those available in Logo. While the language is simple enough for kindergarteners and first-graders to use, it is sophisticated enough for use in solving physics problems at the university level (Thornburg, 1986). The primitives introduced here, however, are sufficient to solve the types of mathematical problems typically explored in the elementary grades. And,

---

[9] Although STOP would also work in this instance, if procedures are being called by other procedures, TOPLEVEL stops *all* procedures, not just the procedure currently being executed.

it is not necessary for the teacher to be a Logo expert to begin to introduce it into the classroom. Many teachers find it very rewarding to discover Logo along with their children. Of course, if you do this, you will wish to have a good Logo reference manual available while you and your pupils explore.

The exercises in Pupil Activity 5.5 (below) are typical of the types of geometry activities children in grades 2 through 6 are asked to do. Typical third- or fourth-graders might spend between 15 and 45 minutes doing *each* of the tasks (of course, some children will be able to complete them at a much faster pace, while others may need considerable help from you or other children to complete them). The activities suggested there should be incorporated into regular mathematics activities with appropriate concrete activities and discussions of the mathematical principles (e.g., degrees, right angles, definitions of shapes, interior and exterior angles, supplementary angles, parallel lines) involved and properly spaced in the curriculum.

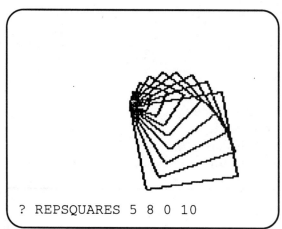

? REPSQUARES 5 8 0 10

Fig. 5.14: Recursive squares

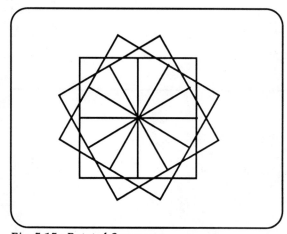

Fig. 5.15: Rotated Squares

Pupil Activities 5.5: Logo Shapes

Objective
Explore angles and polygons
Grouping
Pairs of students
Materials
Logo program packages; computers.
Procedure
The following activities could be completed in several sessions or over a much longer period depending on the level of the learners, the amount of guidance given, and the like. For example, a typical third-grader might complete activity 1 in 10 to 15 minutes, but might spend much more than an hour on activity 12 depending on the number of details in the scene. Other children might complete the same activities in less or more time.
1. Use the FD, BK, RT, and/or LT to draw a right angle on the screen (immediate mode).
2. Draw a diagonal line on the screen.
3. Create and execute a procedure to draw a rectangle on the screen.
4. Create a procedure to draw a rectangle around the edge of the graphics area on the screen.
5. Draw a triangle on the screen.
6. Draw an equilateral triangle; draw an isosceles triangle; draw an obtuse triangle.
7. Draw a regular pentagon.
8. Draw a diamond (a square rotated 45 degrees).
9. Draw 5 separate diamonds on the screen.
10. Draw a circle on the screen.
11. Draw a sequence of squares (triangles, etc.), rotated around a point (Fig. 5.15).
12. Create procedures to draw squares, rectangles, circles and triangles, then put them together in another procedure to create a scene.

Structured Programming in Logo

Structured programming is the art of writing program code that is clear (it is easy to follow the logic of the program), efficient, and easy to debug. Chapter 6 contains a detailed discussion of structured programming. The most important elements of structure in a Logo environment are modularity (each logically distinct task in the program is in its own procedure, which is called from a *master procedure*); internal documentation (comment lines—nonexecuted statements—are used to explain the purpose of the various procedures and blocks of code within procedures); and the use of the structured programming features available in your version of Logo. These techniques are discussed in detail below.

Logo has been touted as a language that promotes structured programming. However, although it is true that Logo promotes modularity, children do not automatically write clear, efficient, well-documented programs in Logo. In fact, a number of aspects of Logo, especially its older versions, tend to encourage programs that are unclear, inefficient, and difficult to read and debug.

Perhaps the aspect of Logo that causes beginning programmers and their teachers the most grief is the fact that Logo statements do not need to be separated by carriage returns (thus do not need to be placed on separate lines). So, it is possible to draw a square as follows:

```
PD
FD 50
RT 90
FD 50
RT 90
FD 50
RT 90
FD 50
RT 90
PU
```

But it is also possible to write the same code like this:

```
PD FD 50 RT 90 FD 50 RT 90 FD 50 RT 90 FD 50 RT 90 PU
```

Or in this way:

```
PD
FD 50 RT 90
FD 50 RT 90
FD 50 RT 90
FD 50 RT 90
PU
```

Although Logo's main building block is the procedure, many children resist using procedures as long as possible, preferring instead to work in the immediate mode despite the difficulty that this causes them. Then, once they are required to use procedures, they will write one long, complex, hard-to-read procedure rather than breaking the problem into logical subunits. Children need guidance in developing appropriate program modules.

Finally, especially in the older versions of Logo, some statements, such as the REPEAT, are hard to read and debug because the entire statement must be completed on one "line"—prior to pressing the RETURN key:

```
REPEAT 10 [MAKE :X 5 PRINT "X IF :Y = 2 [PRINT [NO]].
```

Some of the more recent versions of Logo allow REPEAT statements to be split onto several lines (see below).

Logo teachers can help their students write more structured programs and, hence, be more successful in finding solutions to problems, if they use these guidelines:

1. Move quickly from the immediate mode to the program/procedure mode. Many children will, if you let them, stay in the immediate mode indefinitely. Of course, there are many disadvantages to this, not the least of which is the nonstructured, "seat of the pants," trial-and-error method they use while trying to solve a problem (the opposite of using problem-solving strategies). Another disadvantage is that, after developing some complex set of commands to accomplish a task, that set of commands cannot be re-used or modified—it is lost. Sets of commands that are incorporated into procedures can be re-used as often as desired and can easily be modified. They can be called from other procedures.

2. Insist that the program/problem be divided into sub-tasks and that procedures be written to accomplish them. Also insist that students develop a master procedure that controls the flow of the "program," calling the other procedures.

3. Use comment statements (in most versions of Logo, anything following a semicolon in a line is not executed by the computer) to explain the procedures and tasks in a program.

```
TO COMMENT
;These lines will not be executed by Logo, they
;are only intended to clarify the program.
;This procedure prints "NOTHING"
   PRINT [NOTHING]
END
```

4. Keep a clean, organized workspace—have the students delete procedures not being used and save only those procedures needed to solve the desired problem (not extraneous procedures or procedures left over from the previous problem[s]) (Wiebe & Wiebe, 1989).

5. If your version allows it, have the students organize their Logo code so that it is easy to read: align the contents of procedures, loops, and such, and severely limit the number of commands that are placed on a line. Ask students to list the contents of long REPEAT and IF statements on several lines so that they are easy to read (each substatement in the REPEAT or IF should begin on a new line and be aligned):

```
REPEAT 18 [
   MAKE "NUM :NUM + 5
   PRINT :NUM
```

```
        FD  :NUM
        RT  5]
```

Problem Example 1: Write a program to show that 1257 x 228 can be solved by adding 1257 two hundred and twenty-eight times.

Solution:

```
TO MULTIPLY
 ;Procedure to demonstrate that 1257 times 228 is the
 ;same as adding 228 1257s.

MAKE "TOTAL 0
REPEAT 228 [MAKE "TOTAL :TOTAL + 1257]

 ;***** Print Results *****
 PRINT (SE [The sum of 228 1257s is ] :TOTAL)
 PRINT (SE [1257 times 228 is] 1257*228 )
END
```

Fig. 5.16: Solution to Problem Example 1

Problem Example 2: Find 5 prime numbers larger than 2000.
Solution:

```
TO MAIN
; Main Routine of Program to Determine if Entered Number is
; Prime
;
; ********** Procedures Used in Program ***********
;    CALC   — Determines if divisors exist
;    PRNT   — Prints results
;***************************************************
;
; **************** Variables Used10 ***************
;    NUMBER — Input Amount
; ***************************************************
;
  PRINT [Type a number and this program will tell you]
  PRINT [if it is prime or not.  To stop, type 0.]
  MAKE "NUMBER FIRST REQUEST
  IF :NUMBER = 0 TOPLEVEL ELSE MAKE "FLAG 0
    CALC :NUMBER :FLAG MAIN
END

TO CALC :NUMBER :FLAG
; ***** Procedure to determine if number has a divisor *****
;
```

---

[10] In longer programs, you may wish to have students list and describe the variables they use in each procedure and all procedures used by the program.

```
; ***************** Variables Used *********************
;    FIN    — Square Root of Input Amount (Ends Loop)
;    COUNT  — Loop Control
;*****************************************************************
;
MAKE "FIN INTEGER ( SQRT :NUMBER)
   ; This statement finds the square root of the
   ; inputted number.  When checking to see if a number has
   ; divisors other than 1, you need only check numbers from
   ; 2 to the square root of the number.
   MAKE "COUNT 1
   REPEAT :FIN [MAKE "COUNT :COUNT + 1 IF INTEGER (:NUMBER/
    :COUNT) = (:NUMBER/:COUNT) PRNT 1 STOP]]
   ; This statement checks to see if the number is divisible
   ; by any number from 2 to the square root of the number.
   ; If a divisor is found, the print procedure is called
   ; and the procedure stops (returns to main procedure).
   PRNT 0
   ; If no divisor is found, the print procedure is called.
   ; With the flag set to 0, it prints that the number is
   ; prime.
END

TO PRNT :FLAG
; ***** Procedure to print whether number is prime *****
;
; ************ Variables Used ***********************
; FLAG   — Set to 1 if divisor found, 0 if not found
;*****************************************************************
;
  IF :FLAG = 0 PRINT (SE :NUMBER [ is a prime number.])
    ELSE PRINT (SE :NUMBER [ is NOT a prime number.])
END
```

Fig. 5.17: Solution to Problem Example 2

Logo in the One-Computer Classroom

After completing this chapter, you may ask whether it is possible to incorporate Logo into your classroom if you do not have adequate access to a laboratory full of computers. The answer is, you can! Many teachers successfully integrate Logo into their curriculum with a single computer for an ordinary-sized class. First, they present certain programming concepts to the entire class away from the computer. They may do group activities such as asking pupils to demonstrate what happens when various Logo commands are executed by writing what will appear on the screen on their own paper "screens," by "playing turtle"—moving around a "screen" on the floor to show the turtle movements that will occur when individual commands or procedures are executed, by writing or saying what commands will cause the turtle to move in a desired way, and so on. Subsequently, groups of two through four students are scheduled to work for 15- to 30-minute time blocks at the computer station. Peer tutoring, posters, and activity cards direct the pupils' explorations and assist them when problems occur.

Bibliography

Battista, M. T. and Clements, D. H. 1988. A case for a Logo-based elementary school geometry curriculum. *Arithmetic Teacher* 36(3, November): 11-17.

Billstein, R. and Lott, J.W. 1986. The turtle deserves a star. *Arithmetic Teacher* 33(7, March): 14-16.

Binswanger, R. 1988. Discovering division with Logo. *Arithmetic Teacher* 36(4, December): 44-49.

———.1988. Discovering perimeter and area with Logo. *Arithmetic Teacher* 36(1, September): 18-24.

Bitter, G. G. and Edwards, N. T. 1989 Teaching Mathematics with Technology: Finding number patterns. *Arithmetic Teacher* 37(4, December): 52-54.

Bright, G. W. Teaching mathematics with technology: Logo and geometry. *Arithmetic Teacher* 36(5, January): 32-34.

Campbell, P. F. 1988. Microcomputers in the primary mathematics classroom. *Arithmetic Teacher* 35(6, February): 22-30.

Ernest, P. (1988) What's the use of Logo? *Mathematics in School.* 7 (1, January): 16-20.

Heller, R. S., Martin, C. D. and Wright, J. L. 1985. *Logoworlds.* Rockville, Md.: Computer Science Press.

Jensen, R. J. 1988. Teaching mathematics with technology: Ratios. *Arithmetic Teacher.* 35(8, April) 60-62.

———.1988. Teaching mathematics with technology: Scale drawings. *Arithmetic Teacher* 35(9, May): 36-38.

Newton, J. E. 1988. From pattern-block play to Logo programming. *Arithmetic Teacher.* 35(9, May): 6-9.

Papert, S. 1980. *Mindstorms: Children, Computers, and Powerful Ideas.* New York: Basic Books.

Pattis, R. E. 1981. *Karel the Robot*. New York: Wiley.

Thornburg, D. A. 1986. A computer language at the crossroads: Logo. *A: The Independent Guide to Logo Computing*. 4 (3, March): 78-80, 82, 84.

Watt, D. 1983. *Learning with Logo*. New York: McGraw-Hill.

Wiebe, J. H. 1991. At-computer programming success of third-grade students. *Journal of Research on Computing in Education*. 24(2, Winter): 214-279.

Wiebe, J. H. & Wiebe, M. L. 1989. Teaching Logo systems commands with a cardboard computer. *Logo Exchange*. 7(8, May): 14-16.

Discussion Questions

1. What features of programming languages are used in applications tools such as word processors and spreadsheets?

2. Discuss the importance of teaching children the standard order of operations in mathematical expressions in today's technological world. What kinds of computer activities could help children learn the order of operations?

3. Read the articles by Ernest, Campbell and Keane at the end of this chapter. Next, consider the following: Some critics say that we should not waste classroom time on Logo programming since (1) programming ability is of very little use in the real world, (2) the math concepts learned when programming in Logo could be better learned directly (e.g., use protractors rather than Turtle rotations to learn about degrees), and (3) we should focus on teaching pencil-and-paper computation in the elementary mathematics classroom rather than teaching children to use computers to think for them. Do you agree or disagree with these statements? Why?

Activities

Activity 5.1
Objective
Write Logo statements or procedures to solve problems.
Rationale
Learning a programming language will expand your understanding of computers and will get you started in incorporating programming activities into your math instruction.
Materials
Computer, Logo.
Advance Preparation
Read the appropriate sections of this Chapter carefully.
Procedure
Do the programming activities suggested in the Pupil Activities in this chapter.

Activities 5.2-5.5

Objective
Write structured Logo programs to solve problems.

Rationale
Learning a programming language will expand your understanding of computers and will get you started in incorporating programming activities into your math instruction.

Materials
Computer, Logo.

Advance Preparation:
Read the appropriate sections of this chapter carefully.

Procedure
Write *structured* Logo programs to solve the following:

1. The program accepts as input the lengths of the two legs of a right triangle and prints out the length of the hypotenuse. (The Pythagorean Theorem states that the sum of the squares of the legs of a right triangle—those adjacent to the right angle—is equal to the square of the hypotenuse.)

2. The program accepts as input a dividend and a divisor, then uses repeated subtraction to determine the quotient and remainder. It then prints the quotient and remainder.

3. Write a program similar to problem 2 in Pupil Activity 5.4 above except that the user inputs a number (e.g., 23) and a power (e.g., 5). The procedure calculates and prints the product of the first number and 10 to the power of the second number (e.g., 23×10^5). Call this procedure from another procedure that repeats it until the user types "STOP."

4. Create procedures to draw appropriately sized rectangles, triangle(s), circle(s), and the like. Call them up from a "main" procedure to create a scene with a house, a tree, a sun, etc.

What's the Use of LOGO?

by Paul Ernest

The computer language LOGO is now available in one form or another for most microcomputers. Great claims have been made for the language. LOGO enthusiasts argue that it offers not only a new experience in programming but a revolutionary new way of learning mathematics. How are mathematics teachers to evaluate such claims? In this article I will try to offer a more sober judgement on what LOGO offers teachers and students of mathematics.

The programming language LOGO was created twenty years ago in Massachusetts by a team including its best-known proponent, Seymour Papert. In his book *Mindstorms,* Papert[1] makes some sweeping (and inspirational) claims for one aspect of LOGO, Turtle Geometry. Although LOGO has other uses such as list processing and numerical programming, I shall focus on Turtle Geometry, since the strongest claims are made for this aspect. Papert has the following to say:

> We are learning how to make computers with which children love to communicate. When this communication occurs, children learn mathematics as a living language. Moreover, mathematical communication and alphabetic communication are thereby both transformed from the alien and therefore difficult things they are for most children into natural and therefore easy ones. The idea of 'talking mathematics' to a computer can be generalized to a view of learning mathematics in 'Mathland'; that is to say, in a context which is to learning mathematics which living in France is to learning French.
>
> The kind of mathematics foisted on children in schools is not meaningful, fun, or even very useful.
>
> In order to break this vicious circle I shall lead the reader into a new area of mathematics, Turtle Geometry, that my colleagues and I have created as a better, more meaningful first area of formal mathematics for children. [2]

These are bold and sweeping claims, so bold that they have the ring of an ideological and near-religious fervor. Papert is a visionary who looks ahead to a golden time when all mathematics teachers share his vision and the lack of resources and external examination syllabuses no longer constrain mathematics teaching. We need visionaries, but the skeptical teacher may well ask what LOGO can offer mathematics learners in the here and now. I believe that exposure to LOGO can help children's learning of mathematics in three ways, in both primary and secondary schools.

(a) Logo can help children to learn mathematics content, the concepts and skills of mathematics.

(b) Logo can help children learn the processes of mathematics, particularly the general strategies of problem solving.

(c) Logo encourages new learning and teaching styles, including cooperative group work, discussion and investigations.

Children Learn Mathematics Content through LOGO

In Turtle Geometry children make drawings by directing the Turtle (represented by a small arrowhead, usually) around the monitor screen. This provides concrete experiences of a number of mathematical concepts and skills from geometry and other parts of mathematics.

Estimating Distance

The first problem a child encounters on learning to direct the Turtle's movements is having to decide how many units of forward movement are required to get the Turtle to its desired destination. This is usually tackled by a trial and error process of "homing-in" on the desired endpoint. As children practice drawing with the Turtle on the screen they develop the ability to estimate screen distances in terms of the arbitrary units of length travelled by the Turtle. A number of relationships involving length are often discovered, such as the fact that in a right-angled triangle with sides A, O and H, H is longer than O and A but shorter than O + A.

Angle

The first two concepts a child encounters in controlling the Turtle are those of distance and angle. Use of the commands Left and Right require the child to experiment with and master angle measure in degrees, probably starting with 90° angles. Thus the child uses angle measure and develops angle estimation skills, as with distance. Further, an underlying dynamic conception of angle is developed. For in Turtle Geometry angle is given a concrete meaning as an amount of turn.

In addition to the development of angle concepts and skills, the child discovers the angle properties of plane figures. Thus a straight line angle is 180°, the exterior angle of an equilateral triangle is 120°, and the exterior angles of a polygon sum to 360°. This last fact Papert calls the Total Turtle Trip Theorem: in any Turtle trip which starts and finishes with the same position and heading the Turtle has turned through (a multiple of) 360°.

Shape

In drawing figures with the Turtle the child is exploring the world of shape. Although most children can recognize a square, drawing one with a sequence of moves like the following deepens a child's understanding of the concept.

```
FD 100  RT 90
FD 100  RT 90
FD 100  RT 90
FD 100  RT 90
```

In creating this sequence of instructions the child has realized that the construction of a square requires four right angles, four equal sides, and that there is a four-fold repetition in the construction. Understanding this enriches the web of properties and associations of the concept of square, and lays the ground for the recognition of its fourfold symmetry. The same holds true for other shapes, such as triangles, pentagons and hexagons. Beyond the realm of polygons, the attempt to draw a circle leads to a fuller understanding of that shape too, as a shape which can be approximated by a regular n-gon, for large n.

Symmetry Transformations

The exploration of Turtle Geometry usually begins with the drawing of shapes or figures by means of sequences of instructions to the Turtle, or later, by means of procedures built from these sequences. Once a shape has been designed, it is natural to make copies of the shape on the screen by moving the Turtle to a new position. If the direction of the Turtle is changed, the result is a rotation. If the location of the Turtle is changed, but not its direction, then the result is a translation. Thus two of the three basic symmetry transformations arise simply and naturally in LOGO, as Figure 1 illustrates.

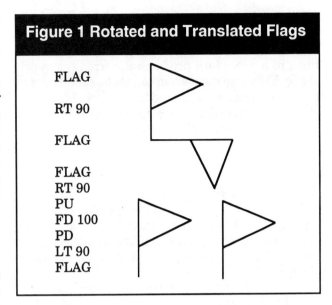

Figure 1 Rotated and Translated Flags

In Figure 1 a rotation and a translation are applied to a simple flag, which is generated by the following procedure:

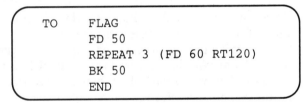

```
TO     FLAG
       FD 50
       REPEAT 3 (FD 60 RT120)
       BK 50
       END
```

The exploration of Turtle Geometry leads to a further development of the notions of translation and rotation. The drawing of directed lines by moving the Turtle provides basis for the concept of a vector. Through experimentation children quickly discover the beautiful, symmetric pattens that arise from the repeated rotations of even the simplest shapes, as in Figure 2. Thus LOGO enables children discover and apply symmetry transformations in a creative and original way.

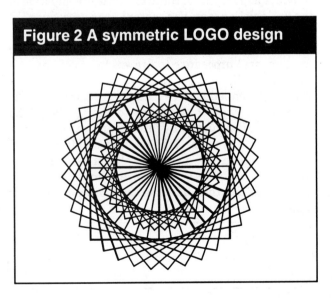

Figure 2 A symmetric LOGO design

The third symmetry transformation, reflection, is not dealt with so easily. Children do discover that the interchange of just the LEFT and RIGHT instructions in a procedure produces a mirror symmetric shape. This transformation cannot be carried out as immediately as rotation and translation. It requires either the adaptation of a procedure, or can be carried out by software written for the tasks.[3]

Enlargement and Similarity

In making screen drawings children often wish to use several sizes of a single shape, for example a square. After keying in instructions for the drawing of different sized squares, children are usually very receptive to the idea of a general procedure, such as the following:

```
TO SQUARE     :SIZE
REPEAT 4 [FD :SIZE RT 9 ]
END
```

Procedures such as this allow a child to generate a family of similar shapes. In creating these shapes the child is coming to grips with the concepts of enlargement, similarly scalefactor and ratio, albeit in concrete form. These are central mathematical concepts of enlargement, which need to be experienced concretely, as in Turtle Geometry.

Variables and Algebra

Algebra is traditionally one of the most difficult areas of mathematics because of its abstraction and formality. Variable is perhaps the central concept of algebra, but unfortunately, as Kuchemann[4] has shown, only a small proportion of children achieve a full understanding of it by the age of 15. LOGO provides a meaningful context for the introduction of the variable concept. For example, the procedure TO SQUARE :SIZE listed above draws squares of any size by employing a variable :SIZE. Here the desire for a general procedure to draw squares of any size gave rise to the need to use a variable. Further, the variable is embedded in a meaningful context, and is named in a way that relates to its meaning. The power that is achieved by introducing variables provides a strong motivating factor for children.

We have seen how a variable allows a single procedure to draw squares of any size. Similarly, the use of a variable in the following short procedure draws a spiral:

```
TO SPIRAL :S
REPEAT 100 [FD :S RT 90 MAKE S :S + 5]
END
```

LOGO is not unique in providing a context in which variables are meaningful. Any programming activity provides such a context, as Fletcher[5] and Booth[6] confirm. However, Turtle Geometry may provide the quickest and least formal entry into programming situations in which the use of variables gives the user great power. Despite this ease of access, one major obstacle to children's easy use of LOGO variables should be mentioned: the horrible syntax involving quotes and colons, illustrated in the procedure SPIRAL.

Recursion

One of the powerful ideas implemented in LOGO is that of recursion, which allows procedures to call themselves up as subprocedures. Children use this idea as a means of initiating repetitive or iterative procedures. Thus, for example, a simple recursive procedure for drawing an enlarging nest of squares is as follows:

```
TO NEST :SIDE
SQUARE :SIDE
MAKE "SIDE :SIDE + 5
NEST :SIDE
END
```

Readers are invited to provide their own illustrations here, by trying out this procedure themselves (with some suitable choice of value for the variable :SIDE).

Recursion is not one of the topics traditionally included in the school mathematics curriculum. However, the concept of recursion does underpin a number of concepts which are common in high school mathematics. These include iteration and iterative processes; inductive and recursive definitions such as those of the factorial, power and Fibonacci funtions and the Euclidean Algorithm; and proof by mathematical induction. As programming and computers increase their impact on the mathematics curriculum it is to be expected that recursion will be given more prominence in the mathematics curriculum, as is already happening with iteration.

Beyond the current mathematics curriculum the idea of recursion leads to a number of exciting areas of mathematics. The self-calling up of procedures is very close to the idea of self-reference. This idea has played a very important role in mathematical logic, from Russell's Paradox to Gödel's Theorems, and beyond. In his exciting book "Gödel, Escher, Bach" Douglas Hofsteder uses the theme of self-reference to link the music of Bach and the art of Escher with Gödel's Theorem.

By way of endless recurrence the idea of recursion leads to the concept of infinity. Recursion in Turtle

Geometry also leads to Fractals, the self-similar curves of Mandelbrot, which in turn lead to the idea of infinitesimals. Recursion in Turtle Geometry also allows the construction of space-filling curves, which played such an important part in analysis in the nineteenth century.

Overall, it has been argued that LOGO, and in particular Turtle Geometry, can help in the learning of the content of mathematics, from the estimation of distance, angle and shape for younger children to ideas including recursion, iteration, self-reference, infinity and infinitesimals, fractals and space-filling curves, for older students. In addition to this range of content and skills, LOGO has a further benefit. Concepts and skills learned during LOGO programming are dynamic and hence meaningful. Consider angle. This is experienced dynamically as an amount of turn. A dynamic conception of angle is more likely to be applied correctly than the more static conception which some children acquire from protractor use. Similarly the notion of the exterior angle sum of a polygon is acquired dynamically in LOGO, as the total amount of turn of the Turtle. LOGO has the virtue of presenting many of the concepts of mathematics dynamically and concretely, that is in a form readily assimilated by many children.

LOGO Teaches General Problem-solving Strategies

There are a number of ways in which programming Turtle Geometry in LOGO teaches general problem-solving strategies.

(i) Children programming in LOGO learn to think algorithmically. This means that children learn to build up sequences of moves to achieve specific goals, whether their own or tasks provided by the teacher.

(ii) Children learn to think procedurally. That is, to "break large problems down into small manageable units—into 'mind-size' bites."[7] Procedural thinking allows children to determine subgoals on the way to a larger goal, and achieve these one at a time.

(iii) Programming on LOGO can provide experience of both "top-down" and "bottom-up" strategies in solving problems. A top-down strategy is like that employed in procedural thinking. A problem or goal is analyzed into a number of modules or subgoals. Each of these is further analyzed until all the subgoals are attainable. Procedures for the subgoals are then put together until the overall goal is achieved. A bottom-up strategy begins with sequences of moves or with existing procedures. These are supplemented and built upon until the goal is achieved. The top-down and bottom-up strategies are basically analytic and synthetic thought in a new guise.

(iv) Programming in LOGO is an activity which parallels all the stages of problem solving in George Polva's heuristics[8]. The analogy can be shown like this[9]

Problem Solving	Programming
understand problem	analyze problem
devise plan	write program
execute plan	run program
check solution and review plan	debug and modify program

In problem solving, after the problem is understood a plan is devised which may be analytic or synthetic. Likewise in LOGO the goal or problem is analyzed and can then be approached by a top-down or a bottom-up programming strategy. In both problem solving and programming the proposed path to the solution, be it a plan or a program, is tried out. Finally the success of this path is evaluated and any flaws in the solution or bugs in the program are remedied.

Evidently there is a very striking analogy between programming in LOGO and problem solving, and both activities involve the development and use of the same general strategies. The same strategies are also evident in mathematical modelling, which involves testing a proposed solution against the situation it is intended to model. This parallels the debugging aspect of programming which is sometimes omitted in problem solving (i.e. not checking solutions), but which is not easily left out in programming.

It has been shown that programming in LOGO teaches and fosters general problem-solving strategies. However, these are not the product of a short-term exposure to LOGO, but require a fairly sustained experience of programming. Furthermore this benefit is not exclusive to programming in LOGO. Although entry to BASIC is not gained so easily as to LOGO via Turtle Geometry, probably the same general problem-solving strategies can be learned, ultimately.

LOGO Encourages New Learning and Teaching Styles

Since the publication of the Cockcroft Report in 1982 teachers throughout the education system have been more aware of the range of styles required for good mathematics teaching. In addition to exposition by the teacher and consolidation and practice, children need to experience investigational work, practical work, problem solving and discussion in their learning. The inclusion of these newer elements in the classroom has been accelerated in secondary schools by the development of the GCSE Examination in Mathematics with its emphasis on investigations, problem solving and practical work in the coursework projects for assessment.

LOGO programming by children is one way in which some of these needs can be met. In the previous section we saw how LOGO helps children to acquire and develop the general strategies and modes of thought which underlie problem solving. Some further aspects are as follows.

Investigations

Investigational work depends on problem solving, and as we have seen, LOGO can offer a great deal here. Beyond problem solving, investigations require children to explore open-ended situations and to set their own goals. Turtle Geometry provides an open, miniature world in which children set their own goals and explore for themselves. Exploration is a very appropriate term to apply to LOGO, for children can not only explore the VDU screen with the Turtle but also the hidden space behind the screen. Beyond this realm of physical exploration LOGO opens up worlds of shape, pattern, design, symmetry and more for exploration.

One of the key features of LOGO for investigational work is that children can set their own goals and pose their own problems. Thus children can feel that they "own" their mathematics in that the goals and approaches are their own creation. In other words, by facilitating problem posing as well as problem solving (and freer exploration) children achieve an ownership over their mathematics. A factor which contributes to this is what has been termed "degoaling." This describes children changing their overall goal then, halfway there, an accidental effect leads them to think of an alternative goal or problem. This type of divergent thought is an asset in investigational work and indeed, in life in general. LOGO provides opportunity for this kind of creative thought which is not, otherwise, common in the mathematics classroom.

Discussion

Genuine pupil-to-pupil and pupil-to-teacher discussion about mathematics in the classroom is not as common as Cockcroft suggests it should be. Since it is common for children to work on LOGO projects in small cooperative groups, a large amount of real discussion takes place. Celia Hoyles and her colleagues in London have recorded a great deal of rich mathematical discussion between pupils working on LOGO.

LOGO work also encourages pupil-to-teacher discussion, and often brings about a new relationship with the teacher as facilitator, collaborator and even co-explorer. Thus LOGO encourages discussion in mathematics.

Personal Qualities

As is the case with other investigational work in mathematics, LOGO programming can have a very positive effect on the personal qualities of pupils over an extended period of time. The achievement of mastery over the Turtle brings confidence. The perception of errors in programming as "bugs" which are to be expected and "debugged" also adds to confidence by diminishing the fear of failure. The setting of own goals and the ability to vary these (in "degoaling") further adds to confidence as children learn that their own judgement is valuable in mathematics lessons. Involvement in LOGO projects with a growing sense of confidence develops persistence in children. Finally, provided that LOGO work is seen as mathematics, it should add to children's interest in mathematics and enhance their attitudes towards mathematics.

Learning

LOGO and Turtle Geometry can provide a dynamic and active learning experience in mathematics. This has been demonstrated for a number of concepts and skills in mathematics as well as for the general strategies of problem solving and investigational work.

Seymour Papert[1] claims that the active learning brought about through LOGO is based on the theories of Piaget. However there is an even closer link with the work of Jerome Bruner. Bruner suggests that our learning is represented mentally in three modes of progressive complexity. These are: *enactive* representation (the memory of an active physical experience such as tying a shoe-lace); *iconic* representation (a simplified pictorial image, such as in a Russian icon) and *symbolic* representation. The learning of LOGO fits neatly into these stages. First of all, young children are best introduced to Turtle Geometry by acting out the movements of the Turtle and then by controlling the physical movements of a robotic Turtle (or the Big Trak toy). Second, children direct the pictorial Turtle around the VDU screen in direct drive. Both these stages lead to iconic representations in their minds. Third, children learn to operate purely symbolically as they write procedures in LOGO. Few areas of mathematics operate so naturally in all these three modes of representation. Even fewer manage to keep these three levels linked so naturally, thus providing meaning at the symbolic level in terms of imagery and action.

Conclusion: LOGO is worthwhile!

I began this article by asking if Seymour Papert's sweeping claims for LOGO were justified. I have shown that one aspect of LOGO programming, namely Turtle Geometry, has a great deal to offer mathematics teaching and learning. It can help in teaching a wide range of mathematical content. It helps develop general problem-solving strategies. It can help to encourage new approaches to the teaching and learning of mathematics. Thus Papert's claims, although extreme, should not be dismissed in their entirety. LOGO is not, of course, a universal panacea. Learners will only get out of it as much as they put in. However, children seem to really enjoy working with LOGO. Even the very least able children with short attention spans will immerse themselves in Turtle Geometry without a break for 80 minutes. For this reason alone, LOGO is a valuable asset to mathematics teaching.

References

1. S. Papert, *Mindstorms: Children, Computers and Powerful Ideas,* Basic Books, New York, 1980.

2. *Mindstorms* pp. 50, 51, 6.

3. R. Goldstein, Mathematics after Logo", *Mathematics Teaching,* 115 (June 1986), 14-15.

4. K. Hart, *Children's Understanding of Mathematics 11-16,* John Murray, London, 1981.

5. T. J. Fletcher, *Microcomputers and Mathematics in Schools,* DES, 1983 (para. 104).

6. L. Booth, *Algebra: Children's strategies and errors,* NFER-Nelson, Windsor, 1984.

7. F. Rezanson & S. Dawson, "The Logo Cult", *Mathematics Teaching* 115 (March 1985), 5-7.

8. G. Polya, *How to Solve it,* Princeton University Press, Princeton, New Jersey, 1945.

9. R. Noss, "Doing Maths While Learning Logo", *Mathematics Teaching,* 104 (Sept. 1983).

Microcomputers in the Primary Mathematics Classroom

by Patricia F. Campbell

Picture a school-board meeting or a meeting of a school district's elementary curriculum committee. Raise the issue of integrating microcomputers into the elementary school's mathematics curriculum, and a debate will ensue. Focus the discussion on the use of microcomputers in the primary classroom, and the remarks will become intense and passionate. Although the diversity of comments prompted by such a discussion cannot be anticipated, two views will probably be voiced. Seeking the promise of a supposed competitive edge, one faction will favor microcomputer use while questioning whether the calculator threatens children's learning of the basics, that is, arithmetic. Citing the added danger of producing socially isolated children who are obsessed with the lure of microcomputers, another group will reject any form of technology in the primary classroom.

Research indicates that both these fears regarding technology are unfounded (Clements and Nastasi 1985; Fein, Campbell, and Schwartz 1987; Hawkins, Sheingold, Gearhart, and Berger 1982; Hembree 1986; Suydam 1982, 1987). Technological advances will not destroy early and primary education, as technology can and should be in tune with the needs and potential of young children. Presented within an environment that simultaneously supports the active, social, and emotional needs of preschool and primary-aged children, the microcomputer offers a unique approach for enhancing academic and creative development in the classroom—for presenting an environment that fosters thinking, imagining children. This approach also encourages a perception of the primary mathematics curriculum as being a time for investigating relationships and solving problems, not just for perfecting routinized arithmetic skills. This article initially suggests some principles to guide the implementation of technology in the early grades. It then describes how Logo can be used as a problem-solving stimulus in the primary grades.

Principles

Technology should be integrated into the curriculum

Young children should not be exposed to microcomputers so that they can learn to "cope" with technology. Technology should be viewed as a means of supporting and enhancing the goals of a complete primary mathematics curriculum, not as yet one more isolated entity that needs to be fit into an instructional program. Microcomputers will influence the content of the primary mathematics curriculum as they offer more diverse settings for problem solving a exploration. However, concept development, problem solving, and examination of characteristics and relationships—not exposure to technology—should drive this instruction.

Technology does not replace use of manipulatives

Manipulative materials play a crucial role in the learning of mathematics. It is necessary for children to reflect on and to verbalize their perception of the mathematics embodied in hands-on materials and for teachers to help students make the critical connection between materials and symbols. Microcomputer software that stimulates shape recognition, counting, classification, patterning, ordering, and set transformation is ideal for young children. However, activities involving microcomputers should not stand alone. Technology should not replace active instruction and exploration with physical objects.

Technological applications should be developmentally appropriate

Microcomputer software is produced for a commercial market that extends beyond the classroom and into the home. Many programs have features that attract the attention of adults, or are easily managed by adults, but are inappropriate for young children. For example, programs with irrelevant color or sound, with involved graphics, or with cluttered layouts are, at best, distracting and, at worst, incomprehensible to young children. The assumed reading level, the presumed attention span, and the required response routine should be appropriate for the age level of the intended user.

Because young children have limited ability and interest in complex procedures for setting up microcomputer programs, and because classroom teachers do not have the time to be continually restarting software, the best software has simple menus that rapidly load programs that are easy for the children to enter and begin using. Programs for young children should be designed so that irrelevant keys are disabled, that is, nothing will happen if keys that are not to be used are pressed. Feedback routines must be carefully designed. Young children are often frustrated by software that ignores incorrect responses and retains the monitor display without some kind of revision. The children tend either to persist in reentering the same incorrect response or to strike the keyboard randomly. At the same time, erroneous responses should not produce more elaborate or engaging graphic displays than do correct responses. Finally, educational software should permit and maintain teacher-prescribed modifications, such as turning off the sound or limiting the size of presented numbers. Further criteria for evaluating software can be found in *Guidelines for Evaluating Computerized Instructional Materials* (Heck, Johnson, Kansky, and Dennis 1984).

Emphasize experiential activities over lower-level skills

Well-written microcomputer programs can promote experiences in solving problems, in predicting outcomes, in breaking a task into smaller problems for subsequent solution, and in drawing conclusions, or they can enhance drill and practice. All these uses are beneficial, but the use of technology in the primary grades to support lower-level skills should be minimized in favor of applications encouraging concept formation, problem solving, and critical thinking. Further, technology should never be used in the primary grades as a device to develop automaticity in the absence of understanding.

Logo in the Primary Grades

By five years of age, young children have developed preliminary logical knowledge, as well as "first draft knowledge" (Gardner 1983, p. 305) of symbolization in language, pictures, dance, pretend play, and such concrete objects as blocks and clay. With schooling, children learn a written notational system to express their developing concepts of language and mathematics. The microcomputer can be a medium in which young children experiment with their understanding of symbols (Sheingold 1986). The programming language Logo permits children to use and create symbols to explore and extend their (exploring an emerging) knowledge of spatial relationships, geometric properties, and sequential thought, as well as their knowledge of number in terms of length and angle measure. At the same time, Logo serves as a setting in which teachers can model and encourage problem-solving strategies.

Some initial considerations

Many teachers have initial concerns regarding the use of Logo with kindergarten, first-, and second-grade children. These concerns usually involve the use of microcomputer disks and the organization of the class. In reality, the disks are not a problem. Even four-year-old children can be taught how to handle and load disks. Actually, the keyboard can be more troublesome, especially if it has an automatic-repeat feature (i.e., the letter or numeral will continue to reenter as long as the key is held down or depressed). Most kindergarten and many first-grade children strike the keys with too much force, activating the automatic-repeat feature. Although the children will learn the correct touch, it is easier if a Logo program is written to handle this error. Primary children are not particularly efficient at the keyboard. However, Logo does not require continual use of all keys; stickers can be used to highlight the critical keys. Simply write a letter on a small colored sticker and place it on the associated key. This labeling limits the number of keys that the children need to scan. Additional tips for teaching keyboarding to primary children can be found in Mason (1985).

First- and second-grade children can readily learn to work in pairs at the microcomputer. However, kindergarten children have difficulty cooperating on a graphic project. Logo is not particularly conducive to the rapid "turn taking" at the keyboard generally desired by pairs of kindergarten children. The ideal is to have two microcomputers in the kindergarten, positioned so that each child can see the other child's monitor screen. This arrangement allows interaction without conflict. lt is best if the microcomputers are placed near the block corner.

Preliminary concepts

Prior to using the keyboard, young children must first experience a number of preliminary concepts regarding directionality and the giving and following of directions. These include the following:

1. Cursor motion does not simply occur. It is the result of a command.

2. Microcomputer commands must be exact.

3. The order of the commands influences the result.

4. The commands RIGHT and LEFT refer to unique directions from the perspective of the moving object.

5. A command to turn to the right or to the left causes a change in direction but not in distance. That is, it causes rotation about a point but not away from a point.

6. A command to move forward or back causes a change in the distance traversed from a point but not in the heading. That is, a command to move back does not mean to turn 180 degrees and move forward.

The distinction between right and left must be taught to all kindergarten and many first-grade children. One effective method is to announce to the class that you will place a sticker or draw a star on the back of the right hand of every child. These directions must be reinforced on a regular basis with each child. For example, "Margaret, I am going to put this sticker on your *right* hand. . . Margaret, what is the name of the hand that has the sticker?. . . Now Margaret has a sticker on her right hand. What is the name of her other hand? . . . I am going to give Sarah her sticker. . . On what hand should I put Sarah's sticker?" Maintain the use of the right-hand sticker as long as the children need the referent. Do not put identifying marks on both hands, as some kindergarten children will mix up the definition of the marks. Rather, mark the right hand and verbally identify each hand as a right or a left hand. It is important to note verbally the identification of both the right and the left hands; otherwise some children may only learn the meaning of right and will only make right turns in off-computer or Logo activities. Given this support for determining left and right, kindergarten children can begin to complete the simpler off-the-computer activities listed in table 1 (through simple paths and mazes).

When using the off-computer activities, care must be taken that all four basic Logo commands enter the children's repertoire. Some children may always start with the same commands or may avoid the use of the commands for left and back. Also, the concept of maintaining the current heading when walking forward or backward must be reinforced, particularly when the child playing the role of commander is not particularly efficient and the student being com-manded has located the hidden object. Primary children tend to rely on ninety-degree turns, or square corners, in their off-computer motions.

Many primary-aged children have had little experience in associating size with turns or in making a gross distinction between large and small turns. Floor paths or mazes are particularly useful for developing this concept, but the children must have a physical guide for their turns. Figure 1 indicates how sheets of paper that are plain on one side and have an angle guide on the other side can be used to create a paper path or maze on the floor. Initially only include forty-five-degree angle guides; ninety-degree angles are then formed by placing the papers perpendicular to each other. Later the angle guides can be varied to present thirty- and sixty-degree measures. In addition to their use as a guide for spatial orientation, these papers furnish a physical model for investigation of the relationship between two or three consecutive acute-angle turns and a perpendicular.

When the children have a firm grasp of the four commands as a way of describing physical movement and a sense of large and small turns to the left and right, Logo can be introduced on the keyboard. Do not rush to the keyboard setting. Once the keyboard has been introduced, periodically return to more challenging variations of these off-computer activities to augment and reinforce Logo instruction. Additional ideas for off-computer Logo activities can be found in the computer education columns of professional peri-

Table 1 Off-the-Computer Activities

Directed Walks

The teacher assumes the role of the "commander" and as the children move to lunch, recess, or the library, uses the Logo commands to direct them. In particular, use the directions of right and left.

While two children are out of the room, hide an object. The two children return and must try to find the object. The child giving directions must remain stationary and must therefore take the perspective of the moving child. The other children use hot and cold signals as feedback.

Maps and Paths

Children can work together to map the path to the cafeteria, gym, office, or playground and then to use Logo commands to interpret the map. Other groups can follow those Logo commands to verify their accuracy.

Teams of students can create paths on the floor using plain sheets of paper. The papers can be laid perpendicular to each other to form ninety-degree turns. If one side of the paper has a guide for selected angles drawn on it, the children may have a guide for turns and paths that do not always produce square corners (see fig. 1). Each team should write the program to map their path; they can also challenge other teams to program their path and compare results.

Mazes

Mazes can be created by arranging classroom furniture, by positioning wooden blocks, or by placing masking tape on the floor. A commander directs a partner through the maze. Teams of children can create and program their own maze, challenging other teams to solve it. Paper mazes can also be created or solved at seatwork. Pairs of kindergarten children are especially fond of making a maze out of large wooden blocks; they then take turns directing and piloting a stuffed animal mounted on roller skate through the maze. Another kindergarten activity is for toy cars to be directed through a maze drawn on dampened sand in the sand box.

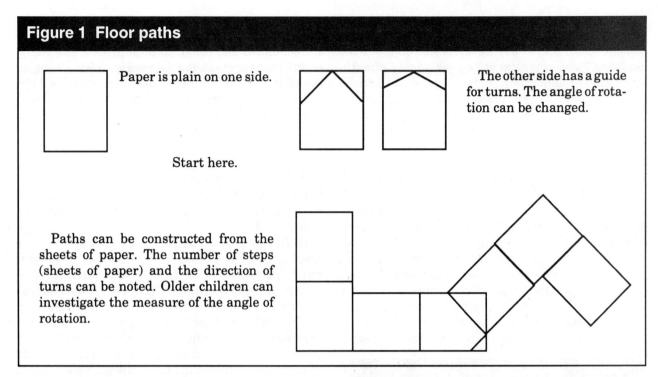

Figure 1 Floor paths

Paper is plain on one side.

The other side has a guide for turns. The angle of rotation can be changed.

Start here.

Paths can be constructed from the sheets of paper. The number of steps (sheets of paper) and the direction of turns can be noted. Older children can investigate the measure of the angle of rotation.

odicals, such as the *Arithmetic Teacher* (e.g., Mason 1985b) or *Teaching and Computers* (e.g., Lough and Tipps 1983a, 1983b; Tipps and Lough 1983).

It is not unusual for some young children to view the microcomputer as a magic box or to sit passively before the microcomputer, waiting for images to appear. The children must learn that, unlike a television, they control the image on the monitor screen and that the microcomputer will only operate if the Logo program, or some other software, is loaded.

Support devices

The transfer from the experienced, three-dimensional world of motion to the remote, two-dimensional plane of the monitor screen can be aided by some support devices. One particularly useful tool is a magnetic board and a triangular piece of cardboard mounted on a small piece of magnetic tape. The cardboard triangle should be marked with an R and an L. Some children also need to see a shaded rectangular region at the base of the cardboard so that the triangle mimics the triangular cursor on the monitor screen (see fig. 2). Commands should first be practiced with the magnetic board flat on the table or floor in front of the children. Then the same commands should be demonstrated with the board held upright, parallel to the monitor screen. Changing planes is not an easy task for some children; the magnetic board permits the children to try out commands on both planes as needed. (Magnetic boards and triangles should be available at every microcomputer.)

Another useful device for each microcomputer is a four-inch by six-inch command card summarizing the direction and distance commands. These should present simple graphic images to define the meaning of the four commands for forward, back, left, and right. Additional commands for naming and saving procedures can be added once the children are more proficient.

Single-keystroke Logo

On the surface, Logo may seem to be nothing more than an electronic crayon, a program that makes it possible to draw graphic images on the monitor screen by moving a triangular cursor. But Logo is far more than a device for immediately executing the four commands for moving the triangular cursor forward, back, left, and right, with helpful hints that tell you when you have a syntax error. It is a procedural language. Therefore, the way to program a solution to a problem in Logo is to break the problem into subproblems, solve the subproblems separately, and then combine all those solutions in some ordered fashion to solve the original problem. This problem-solving strategy is powerful, and in Logo it can be accomplished concretely in terms of graphics.

For example, suppose the problem is "I want to draw a snowman next to an apartment building." This problem can be broken into subproblems: "So I have to draw a snowman, a building, and then move the triangular cursor between them so they line up right." The subproblems can be solved: "I need to figure out how to draw a circle and then stack the circles to make a snowman. I can draw a rectangle and then stack the rectangles to draw an apartment building." Then these solutions are executed in a stated order to solve the original problem: "I'll need to think about where to put the triangular cursor or my snowman could be lying on its side on top of the building." The potential of Logo as a tool for mathematical problem solving is that it permits the child to define and solve subproblems by making up new "procedures" or new "commands" for directing the triangular cursor. In the preceding example, the child may define a procedure for drawing a circle in terms of forward and right moves and then use that circle procedure to define another command for drawing a snowman.

Primary-age children cannot cope with the complexity of the complete Logo language. As noted by Clements (1983-84), to program Logo without frustration requires quite a bit of typing, correct spelling, and proficiency in estimating length and rotation. However, a number of simpler versions of Logo are available. These Instant Logo programs use single keystroke commands to direct the triangular cursor; for example, F moves it forward a fixed number of

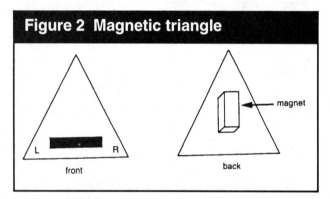

Figure 2   Magnetic triangle

magnet

L   R
front

back

steps and R turns it a fixed number of degrees to the right. Unfortunately, these Instant Logo programs vary widely in their provisions for revision and in their procedural capabilities, that is, in whether the child can make a graphic image by combining pictures that were previously programmed. It is important to note that without these capabilities, an Instant Logo program is no more than an electronic crayon.

The appendix presents a single keystroke Logo for Apple Logo (and Apple Logo II) based on Abelson's (1982) approach. It can readily be adapted for other commercial brands of Logo. To make a copy of this program, you must first load Logo into the microcomputer; then enter the editor and copy all the procedures listed in the appendix; finally, save these procedures as the file SINGLE.KEYSTROKE. The critical differences between this version and other commercial versions of Instant Logo are as follows:

1. In this program, the child must strike the RETURN key after each single keystroke. This feature slows down the execution of the program and in-

creases the probability that the child will observe the graphic impact of the keystroke on the monitor screen. It also allows the program to adjust for the young child's tendency to strike the keys with force, causing multiple entries of a keystroke. Most Instant Logo programs are iterative; an entry of FFFFF will produce a line five times as long as an entry of F. The young child can become confused and lose control of the triangular cursor (or turtle) if he or she thinks an F was struck but the microcomputer is interpreting FFFFF. In this program the user must strike F [RETURN] F [RETURN] F [RETURN] F [RETURN] F [RETURN] to produce a line five times as long as F [RETURN].

2. This program permits the child to revise a graphic image by erasing the effect of the last graphic command (F, B, R, or L) that was entered. This erasure is done by striking E [RETURN]. A sequence of entries such as E [RETURN] E [RETURN] E [RETURN] would erase the effect of the last three graphic commands entered just prior to the Es. In a sense, the E command operates as a pencil eraser. In some versions of Instant Logo, the only way a child revise the image is to erase the screen completely and start over, which is extremely frustrating to the young child. Other versions of Instant Logo permit revision by wiping the monitor screen clean and then redrawing the image, omitting the last command. Young children become confused as to why the image is going and coming; they also may doubt that any revision occurred.

3. In this program the commands for color are not dependent on use of the numeric keys or remembering a numeric code for color. Most Instant Logo programs require a child to look up a color code or to remember that, for example, green is 2 and blue is 5.

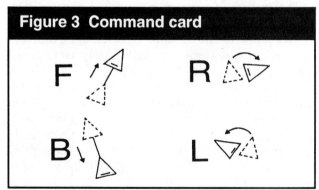

Figure 3   Command card

F   R

B   L

In this program an entry of O [RETURN] causes the child's graphic creation on the monitor screen to freeze. A line is drawn in the lower right-hand corner of the monitor screen. The line changes color as the sequence white, green, violet, orange and blue is gradually recycled. At the bottom of the monitor screen is printed the direction: "PRESS RETURN TO PICK A COLOR." Once the desired color is seen, the child simply strikes the RETURN key. The present color on the color coded line is then fixed, the color coded line disappears, the graphic image is unfrozen, and any subsequent entries of F or B will yield line segments in the selected color.

A brief description of the commands for single-keystroke Logo are presented in table 2.

To use this program, first load Logo into the microcomputer. Then load the file SINGLE.KEYSTROKE. To activate the program, type PRESET. The screen will clear, the triangular cursor will appear in the lower center of the monitor screen, and a rectangular cursor will flash at the lower left. Use the four basic commands (F, B, R, and L) to produce an image on the monitor screen, augmented by commands to change the color (C), to raise or lower the drawing pen (U, D) and to hide or show (H, S) the triangular cursor. Be

Table 2 Definition of Instant Logo Commands

F	FORWARD 10 (steps)
B	BACK 10 (steps)
R	RIGHT 15 (degrees)
L	LEFT 15 (degrees)
U	PENUP
D	PENDOWN
H	HIDETURTLE
S	SHOWTURTLE
C	Selection line cycles through the five available colors. Press RETURN to select a color.
E	Erases the last graphic command (F, B, R, or L)
N	Names all the pictures (procedures) currently in the memory.
M	Makes the current graphic image into a procedure; asks the child to specify a name for the image; adds that name to the listing of procedure names currently in the memory; tells the child the picture is made; clears the monitor screen.
A	Asks the child what picture (procedure) is to be displayed on the monitor screen; lists the available procedures in memory; after the child's response, displays that picture on the screen.
W	Wipes the monitor screen clean; wipes out any graphic commands in the memory that have not already been made into a procedure.

sure to press [RETURN] after every keystroke.

Once a desired image is created, enter M (for MAKE). When asked what the picture should be called or named, respond with the name for the image. Wait for a response stating that your procedure is made. You can then recreate that image on the screen at any time by entering the command A (for ASK). The command A will cause the display to ask you to enter the name of the desired picture, after reminding you what images are available. After you identify the desired picture, the Instant Logo program will execute the procedure associated with that name, producing the image on the monitor screen. The monitor screen does not clear when the command A is executed; this feature permits the simultaneous graphic display of many different procedures or images (solutions to subproblems) on the monitor screen. By moving the triangular cursor on the screen between executions of the command A, the placement of the graphics and the ordering of the execution of the procedures can be controlled. In this way, procedures that produce images can be combined as components of a larger superprocedure, yielding a program to display a picture containing all the images. If you should forget what names you have given your procedures and do not want the microcomputer to draw a picture at that time, enter N and the procedure names will be listed.

Once the desired multiple image (the solution to the entire problem) is complete, enter M to define a superprocedure that calls all the other procedures in the specified order and positions them in the specified locations. To see execution of the complete multiple image, enter A and ask for the superprocedure.

To save this work onto a disk (it is now in the memory of the microcomputer, but it is not on the disk), you must first leave the SINGLE.KEYSTROKE program and return to standard Logo. If you are using an Apple microcomputer, strike CONTROL-G (for Apple Logo) or OPEN-APPLE-ESC (for Apple Logo II); then enter ERASE.INSTANT followed by the usual Logo command for saving files. If you should return to standard Logo and then decide you want to continue working in the single-keystroke program, you can reenter the program by typing SETUP. This command restarts the single-keystroke program, but unlike PRESET, it does not erase all the existing student-created single keystroke procedures from the memory.

Some suggestions for teaching single-keystroke Logo

The four commands for movement (F, B, R, and L) may be presented simultaneously, as the children have already experienced these with their body movements. Use the magnet board as a support device. Evidence shows that although children understand

Figure 4 Two versions of stairs

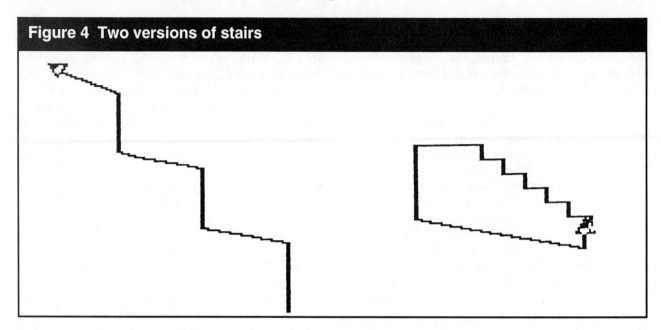

all four commands, they may tend to rely on the forward (F) and right (R) movements (Campbell, Fein, Scholnick, Schwartz, and Frank 1986). The command to move backward (B) may be used as a device for compensating for overshooting or as a means of checking if a desired location has been reached before it is used as a command for drawing. As soon as the children begin to draw on the screen, they will need to use the erase (E) command to remove entries. Some children will use E to change a distance movement (F or B) but will not use it to change rotations. These children will compensate for an overrotation to the right by continuing to turn to the right until a complete revolution is made. Then they will attempt their turn once more with more

precision. Do not try to inhibit these behaviors. The key is to allow the children to gain confidence in their ability to control the turtle or triangular cursor.

As the children become more proficient, the commands to hide or show the turtle and to raise and lower the pen may be introduced, followed by the commands for color. Some children will ask if these commands are possible because they want to use them in their projects. The children may eventually begin to explore the concept of repeating a sequence of commands by recalling their display over and over through the A command. This repetition will lead to a modified version of the standard Logo command REPEAT. Once the children are proficient with the commands, it is possible to place more emphasis on

Figure 5 Rotating Figures

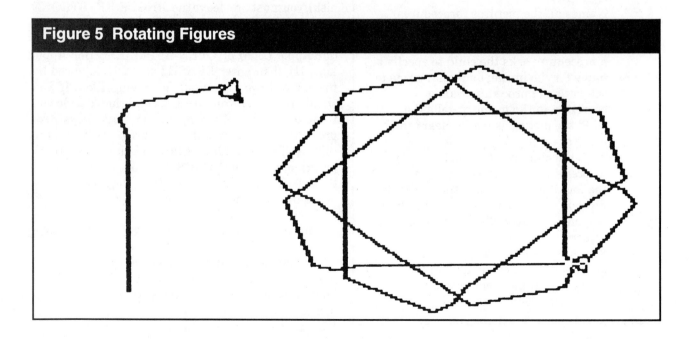

the mathematics of geometric displays and the procedural strategies for defining a display. Many second-grade and most third-grade children will gain enough proficiency in single-keystroke Logo to advance to standard Logo, with its requirement of typing of full command names and estimation of length and angle measure. The support program TEACH allows a child to define a procedure without exiting from the graphic screen. A complete description of TEACH can be found in Clements (1983-84).

When presenting Logo to children in the primary grades, it is important to maintain a balance between classtime spent (a) in teacher-directed instruction presenting aspects of the language or considering the mathematics of a display, (b) in group problem solving and physical manipulations off of the microcomputer, and (c) in completing student projects. Do not limit instruction to the completion of commercially prepared activity cards. Student projects may sometimes be defined by the teacher, but time should also be allotted for student-determined, teacher-approved projects. It is critical that primary children have time to create their own displays; the key is to balance the children's exploration with an expectation that their work will show growth. Always, however, take time to discuss and share each child's work.

For example, figure 4 displays a kindergarten child's two images for stairs. Two images appear because the child was not satisfied with the initial image, spent some directed time exploring the characteristics of stairs, and then produced the second image. What kind of mathematical learning could programming stairs promote? To begin with, the kindergarten class could investigate the characteristics of some stairwells in their school. This investigation may lead to the determination that the steps, when viewed from the side, do not simply turn and go up and down but have square corners that can be "filled" with upright textbooks, sheets of paper, or cardboard rectangles. The children may then move to exploring the construction of square corners with crayons and paper, footsteps on floor tiles, and floor paths before making square corners with Instant Logo. Similarly the width and height of a number of the steps could be measured and compared with nonstandard units (e.g., wooden blocks), yielding consistent measures for height and for width. Following offcomputer transfer to drawings, paths, or footsteps, the children may again use Instant Logo to program stairs. Subsequent class discussion may involve counting and comparing the number of rotation commands that the children entered to produce square corners in their stairs, as well as the number of distance commands in their procedures.

Figure 5 displays another child's study of figural rotation. The drawing on the right is produced by rotating the "stem figure" on the left. This work was also completed by a kindergarten child who was beginning to learn how to make and call back procedures, yielding a superprocedure. The instruction that preceded this construction was not simply the mechanics of making a procedure for the stem figure (via the command M) and repeatedly calling back the procedure (via the command A). Rather the children first used pattern blocks to investigate rotating figures in three-dimensional space. Transfer to the microcomputer screen yielded the study of the influence of the initial heading of the triangular cursor each time the procedure was recalled on the inclination of the stem figure. This consideration led to an off-computer investigation of symmetry. At the end of the school year, the children had decided that the heading of the cursor determined how many times you had to recall the stem figure before you were "back where you started from." The exact relationship between the angle of that heading and the number of rotated figures was not investigated with these kindergarten children. However, older children could collect those data, make predictions, and hypothesize a relationship.

Summary

Microcomputers permit even young children to investigate mathematical relationships and to solve problems. Certain behaviors seem to occur when these devices are available. Children attempt to define verbally what the problem means to them and then try to figure out a solution. Children learn to act out or model an event to understand and visualize the associated problem. That process is the rationale behind the many off-the-computer activities listed in this article. If a supportive, creative teacher is present, Instant Logo can promote individual investigation and act as a catalyst for mathematical exploration. This kind of activity can be a very exciting experience for young children and their teachers. The challenge is to capitalize on it.

References

Abelson, Harold. *Apple Logo.* Hightstown, N.J.: BYTE/McGraw-Hill, 1982.

Campbell, Patricia F., Greta G. Fein, Ellin K. Scholnick, Shirley S. Schwartz, and Rita E. Frank. "Initial Mastery of the Syntax and Semantics of Logo Positioning Commands." *Journal of Educational Computing Research* 2 (November 1986):357-78.

Clements, Douglas H. *Computers in Early and Primary Education* Englewood Cliffs, N.J.: Prentice-Hall, 1985.

—— "Supporting Young Children's Logo Programming." *The Computing Teacher* 11 (December/January 1983-84):24-30.

Clements, Douglas H., and Bonnie K. Nastasi. "Effects of Computer Environments on Social-Emotional Development: Logo and Computer-Assisted Instruction." In Cleborne D. Maddux (ed.), *Logo in The Schools* pp. 11-31. New York: The Haworth Press, 1985.

Fein, Greta G., Patricia F. Campbell, and Shirley S. Schwartz. "Microcomputers in the Preschool: Effects on Social Participation and Cognitive Play." *Journal of Applied Developmental Psychology* 8 (April-June 1987): 197-208.

Gardner, Howard. *Frames of Mind.* New York: Basic Books, 1983.

Hawkins, Jan, Karen Sheingold, Maryl Gearhart, and Chana Berger. "Microcomputers in Schools: Impact on the Social Life of Elementary Classrooms." *Journal of Applied Developmental Psychology* 3 (October-December 1982):361-73.

Heck, William P., Jerry Johnson, Robert J. Kansky, and J. Richard Dennis. *Guidelines for Evaluating Computerized Instructional Materials.* 2d ed. Reston, Va.: National Council of Teachers of Mathematics, 1984.

Hembree, Ray. "Research Gives Calculators a Green Light." *Arithmetic Teacher* 34 (September 1986):18-21.

Lough, Tom, and Steve Tipps. "Lesson One: Introductory Activities." *Teaching and Computers* 1 (September 1983a):38-39.

——"Lesson Three: More Turtle Control." *Teaching and Computers* 1(November /December 1983b):49-51.

Mason, Margie."Computer Corner: Keyboarding." *Arithmetic Teacher* 33 (November 1985a):51.

——"Computer Corner: Maze Days." *Arithmetic Teacher* 33 (October 1985b):46.

Sheingold, Karen. "The Microcomputer as a Symbolic Medium." In *Young Children and Microcomputers*, edited by Patricia F. Campbell and Greta G. Fein. pp. 25-34. Englewood Cliffs, N.J.: Prentice-Hall, 1986.

Suydam, Marilyn N. "The Use of Calculators in Pre-College Education." Columbus, Ohio: Calculator Information Center, 1982. (ERIC Document Reproduction Service No. ED 220-273).

——"What Are Calculators Good For?" *Arithmetic Teacher* 34 (February 1987):22.

Tipps, Steve, and Tom Lough. "Lesson Two: Grids, Mazes and Maps." *Arithmetic Teacher* 1 (October 1983):51-52.

Appendix

Instant Logo

```
TO CHANGE.COLOR
SETPC :X
SETHEADING 90
FD 30 BK 30 WAIT 100
MAKE "COM1 READKEY
TEST :COM 1 = [ ]
PE FD 30 BK 30 PD
IFF [MAKE "Y :X + 1]
IFF [MAKE " X ( REMAINDER :Y 6)]
IFF [IF :X = O [MAKE " X 1]]
IFF [OUTPUT CHANGE.COLOR]
IFT [OUTPUT :X]
END


TO COLOR
SPLITSCREEN
```

```
PR [ ] PR [ ]
MAKE "HOLD1 POS
MAKE " HOLD2 HEADING
PU HT
SETPOS [105 -75]
PD
MAKE " X 1
PR [PRESS RETURN TO PICK A
    COLOR.]
MAKE " TINT CHANGE.COLOR
PR [] PR []
GOBACK
OUTPUT ( LIST " SETPC :TINT )
END


TO GOBACK
PU
SETPOS :HOLD1
SETHEADING :HOLD2
PD ST
END


TO READKEY
IF KEYP [OUTPUT READLIST]
OUTPUT "
END


TO ASK
SPLITSCREEN
PR[]
OR [WHAT PICTURE DO YOU WANT
    TO SHOW?]
PR SE [YOU HAVE: ] :PICTURES
MAKE " RESPONSE READLIST
IF :RESPONSE = [ ] [SPLITSCREEN
    STOP]
TEST MEMBERP FIRST :RESPONSE
    :PICTURES
IFF [PRINT SE [YOU DON'T HAVE A
    PICTURE CALLED :RESPONSE]]
IFF [WAIT 200 SPLITSCREEN STOP]
FULLSCREEN RUN.AND.RECORD
    :RESPONSE
CLEARTEXT
SPLITSCREEN
END


TO COMMAND
MAKE " COM READLIST
IF :COM = [F] [PR [ ] RUN.AND.
    RECORD [FORWARD 1O]]
IF :COM = [R] [PR [ ] RUN.AND.
    RECORD [RIGHT 15]]
IF .COM = [L] [PR [ ] RUN.AND.
    RECORD [LEFT 15]]
IF :COM = [B] [PR [ ] RUN.AND.
 RECORD [BACK 1O]] IF :COM =
E] [PR [ ] CLEARTEX
   UNDO] IF .COM = [H] [PR [ ]
    N.AND.    REC
RD [HIDETURTLE]] IF :COM = [
] [PR [ ] RU
.AND.    RECORD [SHO
```

```
TURTLE]]
IF :COM = [U] [PR [ ] RUN.AND.
     RECORD [PENUP]]
IF :COM = [D] [PR [ ] RUN.AND.
     RECORD [PENDOWN]]
IF :COM = [C] [RUN.AND.
     RECORD COLOR]
IF :COM = [W] [CLEARSCREEN
     SETUP1 ]
IF :COM = [M] [LEARN]
IF :COM = [A] [ASK]
IF :COM = [N] [SPLITSCREEN PR
     [YOUR PICTURES ARE NAMED]
     RUN.AND.RECORD [PR :PICTURES]]
END


TO PRESET
MAKE "PICTURES [ ]
SETUP
END


TO SETUP
TEXTSCREEN
PRINT [SINGLE KEYSTROKE
     PROGRAM]
PRINT [BY DOUGLAS H. CLEMENTS]
PRINT [PATRICIA F. CAMPBELL]
WAIT 100
MAKE "HISTORY [ ]
SETPC 1 CLEARTEXT
CLEARSCREEN
HT PU BK 50 PD ST
SPLITSCREEN
RECYCLE
SINGLEKEY
END


TO SINGLEKEY
COMMAND
RECYCLE
SINGLEKEY
END


TO UNDO
SPLITSCREEN

PR [ ] PR [ ]
IF:HISTORY = [ ] [STOP]
MAKE "C FIRST LAST :HISTORY
TEST MEMBERP :C [FORWARD BACK
     RIGHT LEFT]
IFT [MAKE "NUM LAST LAST
     :HISTORY]
IF C = "FORWARD [PE BK :NUM PD]
IF C = "BACK [PE FD :NUM PD]

IF C = "RIGHT [LT: NUM]
IF C = "LEFT [RT: NUM]
MAKE "HISTORY BUTLAST :HISTORY
END


TO LEARN
SPLITSCREEN
```

```
PR [ ] PR [ ]
PR [WHAT DO YOU WANT TO CALL
THIS PICTURE?]
MAKE" PROCEDURE.NAME UN IQUE
WAIT 100
DEFINE :PROCEDURE.NAME FPUT [ ]
     :HISTORY
MAKE" PICTURES LPUT
     :PROCEDURE.NAME :PICTURES
FULLSCREEN
CLEARTEXT SPLITSCREEN
PR SENTENCE :PROCEDURE.NAME
     [IS MADE.]
WAIT 600
SETUP1
END


TO UNIQUE
MAKE " TITLE FIRST READLIST
IF :TITLE = [ ] [SPLITSCREEN STOP]
TEST MEMBERP :TITLE :PICTURES
IFF [OUTPUT :TITLE]
IFT [PR [YOU HAVE A PICTURE BY
     THAT NAME.]]
IFT [PR [WHAT DO YOU WANT TO
CALL THIS?]]
IFT [OUTPUT UNIQUE]
END


TO SETUP1
TEXTSCREEN
MAKE "HISTORY [ ]
SETPC 1 CLEARTEXT
CLEARSCREEN
HT PU BK 50 PD ST
SPLITSCREEN
RECYCLE
SINGLEKEY
END


TO RUN.AND.RECORD :ACTION
RUN :ACTION
MAKE " HISTORY LPUT :ACTION
     :HISTORY
END


TO ERASE.INSTANT
ERASE [RUN.AND.RECORD SETUP1
     UNIQUE LEARN UNDO SINGLEKEY
     SETUP PRESET]
ERASE [COMMAND ASK READKEY
     GOBACK]
ERASE [COLOR CHANGE.COLOR]
END
```

Spatial Problem Solving with Logo

by Dorothy L. Keane

It is important to remember that students learn problem solving by solving a variety of problems, not only those that emphasize analytical and arithmetic/algebraic skills, but also those which use and enhance spatial skills. The need to experience problem solving in a variety of forms was reinforced by the research of Davidson (1983). Working with a team of medical experts in neurology, she defined two learning styles in mathematics related to children's brain hemispheric preferences. Students with an aptitude for spatial tasks are generally determined to be right-hemisphere dominant and prefer solving problems that emphasize inductive reasoning. Sometimes they overlook important details when analyzing problems. Those who excel in performing routine sequential tasks that involve logical reasoning are more left-hemisphere dominant. They prefer solving problems that use deductive reasoning. While using a step-by-step approach, they may neglect to see the "whole." After testing over 1200 children ages 5 to 18 to assess ability, achievement and learning style, it was concluded that ". . . successful mathematical experience is predicated upon successful harmony of both learning styles." (Davidson, 1983) Evidence suggesting that spatial ability relates positively to problem-solving performance is frequently noted in research literature (e.g. Guay, & Mc Daniel, 1977; Meyer 1978; Parker; Sodamann, 1991).

The ability to mentally picture and manipulate a problem situation is very important when it comes to solving all kinds of mathematical problems. For example, if a child is able to visualize the people and money in a problem where money changes hands as she reads it, she will likely understand it. This is the first step in the problem-solving process. Or if a student can picture a number line, he will have a tool to help him solve problems involving integers.

One of the branches of mathematics that gives particular opportunities for spatial problem solving is geometry. According to the National Council of Teachers of Mathematics (*Curriculum Evaluation Standards for School Mathematics*, 1989),

> Spatial sense is an intuitive feel for one's surroundings and the objects in them. To develop spatial sense, children must have many experiences that focus on geometric relationships; the direction, orientation, and perspectives of objects in space; the relative shapes and sizes of figures and objects; and how a change in shape relates to a change in size.

Logo, a derivative of the list-processing language LISP, was developed in the 1960s by Seymour Papert and a group working in the Artificial Intelligence Laboratory at Massachusetts Institute of Technology (MIT). It evolved from the premise that programming can be an effective environment for solving interesting problems for students of all ages. It has been taught to children as young as four years old, at all levels of elementary and secondary schools, and is used to solve college physics problems.

The component of Logo which allows students to explore relations and spatial problems is called Turtle Geometry. By entering words that are in the Logo repertoire (commands), children can direct the turtle (a triangle in the center of the screen) to move about the screen, drawing a path as it moves. The movements of the turtle are relative to its position. The command RIGHT 45, for example, will turn the turtle 45 degrees to the right (clockwise) of its current direction (heading). FORWARD 70 commands the turtle to move forward 70 turtle units in the direction it is heading, drawing a path as it moves. This graphic system gives children the opportunity to view the turtle's microworld from its perspective. When deciding how to rotate or move the turtle, attention must be given to its present heading and location on the computer screen, while at the same time visualizing the turtle's movement as result of the proposed commands.

Turtle graphics allow for a large variety of spatial problem-solving activities. It is an excellent medium for exploring direction, orientation, proportion, and so on—as suggested by the NCTM standards. Not only can children explore these concepts within the microworld of the computer screen, teachers can arrange activities so that children translate shapes from one space (e.g., the chalkboard or the floor of the classroom) to another (e.g., the computer screen). In order to solve a particular problem, children may visualize themselves moving and turning so that they can understand how to command the screen turtle to move. Or, students may view a shape on the chalkboard and then develop a set of commands to create that shape on the screen. As children create more complex shapes and patterns, their visualization capabilities should become more sophisticated.

Activities for Spatial Problem Solving in Logo

When elementary and middle school children begin working with Logo, it is helpful to use the analogy of "spoken languages" to help them understand how one communicates with a computer via a programming language. For example, we may ask the students what languages they speak. Then ask a student to tell you to do something in a language other than English. After not doing what she asked you to do, you can discuss the idea that in order to get the computer to do what you want, you must use a language it understands. In this case the language is Logo. There are certain words in the Logo vocabulary that we need to learn so we can teach the computer to do what we wish. You may wish to emphasize that the students teach the computer. This puts them in charge, not the computer (Keane, 1984).

Until students are familiar with "the turtle's view of the world," they should be allowed to experiment in the immediate mode. Initial activities can involve the FORWARD (FD) and RIGHT (RT) command. One of the earliest activities is to ask students to use these two commands to draw a square. First, have them enter:

```
FD 100
```

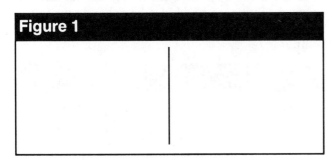

Figure 1

If they want to continue and make a square, students will need to visualize what the square should look like, the present position of the turtle and what commands will have to be given to complete the task of making a square.

```
RT 90
FD 100
```

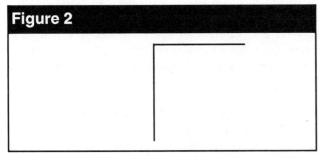

Figure 2

Now that there are two sides to this square, the challenge for the beginner is to try to view the "world"

the way the turtle does. Looking at the computer screen (fig. 2) might lead one to command the turtle to move "down". That word isn't in the Logo vocabulary. Instructions that could be added are:

```
RT 90
FD 100
```

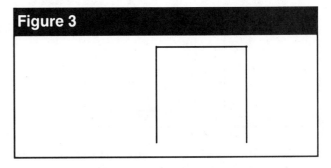

Figure 3

Finally, instructions that will result in a square could be:

```
FD 100 RT 90
FD 100 RT 90
FD 100 RT 90
FD 100 RT 90
```

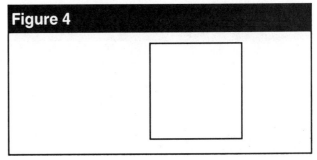

Figure 4

Familiarity with simple commands: FORWARD (FD), BACK (BK), RIGHT (RT), LEFT (LT), PENUP (PU), PENDOWN (PD) can give students many opportunities for meaningful spatial problem solving. It isn't necessary to get students into writing programs (procedures) too soon. There is much to be gained from experimenting and discovering spatial relationships when working in the immediate mode. It is important to give students opportunities to solve problems that use their strength as well as those that address any deficit they may have. Students who are "left brain dominant" may more easily solve problems that ask them to analyze Logo instructions line by line and predict the outcome. Students who are "right brain dominant" may be able to look at a drawing and enter the instructions while mentally visualizing the result. The following activities are designed to address the needs of both learning styles with the hope that students will acquire a successful harmony of both learning styles.

Sample Activities to Discover Spatial Relations in the Turtle's World (Immediate Mode)

Teach the turtle to draw a rectangle (but not a square).

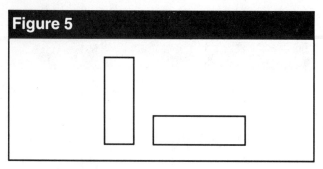

Figure 5

This builds on the "square" instructions. Students need to keep in mind the attributes of a rectangle (a four-sided figure, whose adjacent sides are unequal and intersect at right angles).

• Teach the turtle to draw the skinniest rectangle you can think of.

Students can discover that Logo can handle numbers < 1.

• Teach the turtle to draw an equilateral triangle.

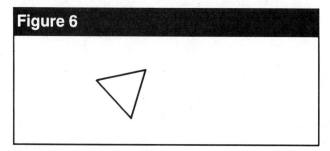

Figure 6

Students who have learned that the interior angles of an equilateral triangle are 60 degrees immediately bring that knowledge to this task. When they find that 60 degrees doesn't work, they have to test a new hypothesis. Some students become frustrated using a "trial and error" approach. In such cases, questions can be asked to lead them to discover. *Sample Questions:* "Since this is an equilateral triangle, what do you know about the angles?"(All three angles are equal)

"When you had the turtle draw a square, what was the measure of each angle? (90°)

"What was the total rotation of the turtle?" (360°)

"Does this knowledge help when drawing the triangle?" (Total turtle trip theorem: The total rotation of the turtle for a closed figure is +/-360°.)

Drawing the equilateral triangle is good problem for introducing or reviewing interior, exterior and supplementary angles.

Ask students to extend a side of the triangle to see

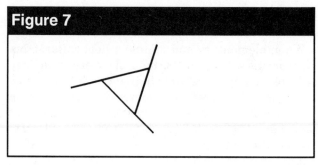

Figure 7

how the turtle rotates. They should discover that the turtle rotates along the exterior angle; the 60° measure that they remember from previous learning refers to the interior angle. And given that the sum of the interior angle and its exterior angle is 180° (supplementary angles) the turtle rotates 120°

• Use what you have learned about triangles to teach the turtle to draw a) A pentagon b) A hexagon c) An octagon with a perimeter of 100 d) A n-gon

• Turn the monitor off and enter the instructions to get the turtle to draw this city skyline (Fig. 8). Keep a mental picture of what the turtle will have to do on the screen as you enter the instructions. Then turn on the monitor to see if the turtle did what you expected. Switch the computer to the text mode (CTRL-T in most versions of Logo) to see the text you entered. (Students could also be asked to write the instructions on paper first and then enter them in the computer.)

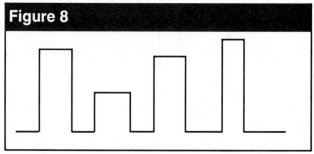

Figure 8

This skyline problem gives students an opportunity to mentally picture what the turtle will need to do. It gives students an opportunity to become aware of their skills in spatial visualization. Most have had little practice. Yet this is an important skill for study in geometry and physics.

• Here's a challenge! Teach the turtle to draw the triangular design shown in figure 9. It is interesting to observe how students approach this problem. You might encourage them to use the triangles as a base. There are many ways to derive this design. Great discussions take place after students have discovered their own solution. If a student used the triangles as a base to solve the problem, then ask her to try the approach of drawing the hexagon and the indicated diagonals.

Figure 9

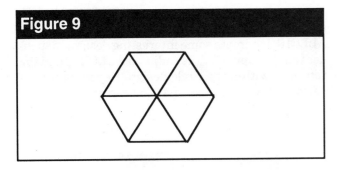

• On a piece of paper, draw the designs that would result from each of the following instructions. Then enter them in the computer and see if your design matches the turtle's. It is also important to ask students to analyze a set of instructions and draw the resulting design. This requires a different kind of thinking where students must interpret text and translate it to the turtle's world using spatial visualization

INSTRUCTIONS 1	INSTRUCTIONS 2
FD 30	FD 50
RT 45	BK 30
FD 30	RT 90
RT 45	FD 50
FD 30	BK 30
RT 45	RT 90
FD 30	FD 50
RT 45	BK 30
FD 30	RT 90
RT 45	FD 50
	BK 30
	RT 90

INSTRUCTIONS 3

```
LT 90
REPEAT 10 [FD 30 BK 30 RT 20]
```

Sample Activities That Use Logo Procedures
• This is a picture of "Harry." On a piece of paper, write the instructions for this "Harry" or your depiction of "Harry." Later you will be able to enter those instructions to see if the turtle draws your Harry.

This task requires planning, spatial visualization

Figure 10

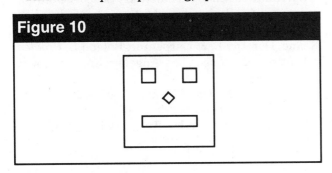

and even some arithmetic computation. You may wish to have students work in pairs in both writing and entering the instructions into the computer. Although it is not essential, you may wish to introduce students to writing procedures when giving this problem. Then they can easily correct any errors in their instructions (debug).

Writing procedures gives students the opportunity to add words to Logo's vocabulary. Once the instructions for the chosen word (procedure name) are entered, that word is added to Logo's repertory (or dictionary). Of course, the procedures need to be saved to disk for long-term storage.

An enjoyable problem is one that asks how to get Logo to draw a circle. (Some versions of Logo such as Terrapin LogoPlus, happily, do not have a built-in command for drawing a circle) Building on experiences with polygons, students can explore the rotations and lengths of the turtle steps that will result in a polygon closely resembling a circle. Most eventually discover that

```
TO CIRCLE
REPEAT 360 [ FD 1 RT 1]
END
```

produces a circle. They find that this is very slow, and always results in a single size circle. Arriving at CIRCLE2 requires a fair amount of thinking. "Why is it faster and still the same size as CIRCLE?"

```
TO CIRCLE2
REPEAT 36 [FD 10 RT 10]
END
```

Next, students may be asked to teach the turtle to draw a circle that would be good for the eye of a rabbit, and one that would be good for a head. This encourages experimentation to find that changing the length of the turtle path is one way of drawing different size circles.

```
TO CIRCLE3
REPEAT 36 [FD 8 RT 10]
END
```

Finally, you may ask what the circumference or total turtle path of CIRCLE3 was (36 x 8 = 288). After learning how to use variables in Logo, students are able to write a simple general procedure that resembles CIRCLE3, which allows them to draw circles of arbitrary size.

```
TO CIRC :S
REPEAT 36 [FD :S RT 10]
END
```

Often it is difficult to find good applications for geometric concepts. After students memorize the formulas and solve algorithms, they feel no real need for the knowledge. Logo gives many opportunities to use mathematical knowledge. The examples of drawing a rabbit or a person or a snowman require specific size circles. Using the formula for circumference: C = 2 x pi x r within procedures adds the feature of being able to specify a specific radius or diameter when calling on the procedures. Here are two examples, one with radius as the input and the other with diameter as the input:

```
TO ALLCIRC :R
HT
REPEAT 36 [FD(2*3.1416 *:R)/36 RT 10]
END

TO ALLCIR :D
HT
REPEAT 36 [FD (3.1416 * :D)/36 RT 10]
END
```

Some fun problems that use these two procedures:
• Use ALLCIRC and ALLCIR to draw two circles of the same size.
• Write a procedure that uses ALLCIRC to draw this: (Fig. 11)

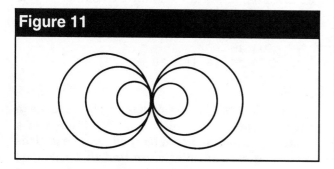

Figure 11

• Use the procedure GROWCIR with different radii to see what happens. Predict the number of circles that will be drawn dependent on the input. Then change the size of the increment and see the resulting design.

```
TO GROWCIR :R
HT
IF :R > 50 THEN STOP
ALLCIRC :R
GROWCIR :R + 10
END
```

• Write a procedure similar to GROWCIR. Call it SHRINKCIR. Have it draw circles that decrease in size.

• Here is a procedure called SPINCIRC. It, too, uses the building block capability of Logo. It uses ALLCIRC to create some interesting designs, dependent on the inputs for the radius of ALLCIRC and the rotations within the circle design. Experiment with various inputs and see what you can create.

```
TO SPINCIRC :R :S
REPEAT :S [ALLCIRC :R RT 360/:S]
END
```

Conclusion

Most educators would agree that students at every grade level need to increase their abilities in problem solving. This will only occur if they are given the opportunity to use their problem-solving skills in a variety of situations. The problems chosen should build on the learning preferences of students and strengthen any learning deficits they may have. Even though Logo is not limited to turtle graphics, exploring in the turtle's microworld can enhance the spatial thinking of students of any age. It is good to keep in mind that using Logo effectively relies on a good teacher. The climate of "exploration" and "invention" does not just happen. It takes an effective teacher to create that climate—to know when to allow students the space and independence they need and when to give assistance. Asking questions rather than telling facts is a beginning. I invite the reader to discover the many other ways that Logo can be used in the mathematics curriculum and to open the door of Logo discovery to their students.

Bibliography

Davidson, Patricia S. *Mathematics Learning Viewed From A Neurobiological Model for Intellectual Functioning*. (A Final Report, Volume One) National Institute of Education Report, NIE-G-79-0089, 1983.

Guay, Roland & McDaniel, Ernest. "The Relationship Between Mathematics Achievement and Spatial Abilities Among Elementary School Children." *Journal for Research in Mathematics Education.* 8,211-215, 1977.

Keane, Dorothy L. & Carpenter, Roland L. *Problem Solving with Logo*. Los Angeles: Kencar Associates, 1984.

Krutetskii, V.A. *The Psychology of Mathematical Abilities in Schoolchildren*. (J. Kilpatrick & I. Wirszup, Eds., J. Teller, Trans.) Chicago: University of Chicago Press,1976.

Meyer, R.A. "Mathematical Problem-solving Performance and Intellectual Abilities of Fourth-grade Children." *Journal for Research in Mathematics Education.* 9,334-348,1978.

National Council of Teachers of Mathematics. Curriculum and Evaluation Standards for School Mathematics. Virginia: NCTM, 1989.

Papert, Seymour. *Mindstorms: Children, Computers and Powerful Ideas.* New York: Basic Books, Inc., 1980.

Parker, Eloise P. "Developmental and Gender Differences in the Use of Imagery in Problem Solving." *Dissertation Abstracts International*, Vol. 52, No. 2, August 1991, p.480a.

Sodamann, Paul E., "Visual Imagery and Achievement Test Outcomes in Seventh Grade Science." Dissertation Abstracts International, Vol. 52, No. 1, July 1991, p. 73a.

Eight Ways to Get Beginners Involved in Programming

by Diane McGrath

Whether you teach third grade or adults, you can teach so that students with lots of different interests will enjoy learning programming.

I used to teach psychology to college freshmen. I don't thing I ever did succeed in finding a textbook that taught things in the right order. A typical approach was to teach about the nervous system, perception, classical conditioning, and on and on until, if we moved quickly enough, during the final week of classes students might learn something about personality and abnormal psychology—the reason most of them took the course in the first place. By that time they didn't care anymore.

Now I teach teachers about computers, and I find that many programming books do the some thing. First there's RAM and ROM and CPU and maybe even binary arithmetic, and then (in BASIC, for example), PRINT, LET, and DATA..READ, with lots of tax and payroll examples. Students wonder what this has to do with anything they can use or do. Much, much later they get to graphics, INPUT, branching, and string manipulations, with which they can do more interesting things. High school textbooks are generally even worse than the books for teachers.

I would like to share with you some ways I have found over the years of getting everyone (well, nearly everyone) involved in programming, right from the beginning. If you wait too long to interest them, you will lose them. My experiences come from teaching people of all ages and types, from 5-year-olds to retirees, from future hackers to accomplished musicians, how to program in Logo, BASIC, and/or Pascal. These experiences are also backed up by research.

People come to programming with different motivations, backgrounds, interests and biases. Programming can and should be taught with this variety in mind. The suggestions below are based on the premise that you should teach the concepts in an order that makes the most sense to beginners, and that is most intrinsically interesting to beginners. The goal is not to weed out those who are not destined to become computer scientists; it is to involve and educate as best we can the greatest number and variety of students.

Making Assignments Motivating to All

1. Start with graphics.

The inventors of Logo designed the language to have a graphics entry point, a motivating and understandable way to learn some powerful programming ideas (Papert, 1980). Introduce graphics as soon as possible—in the first or second class period, even in Pascal. If you let students loose on a graphics project of their own design, they will spend considerably more time perfecting it than they do on other assignments. Graphics is an excellent arena in which to introduce other early programming concepts, such as variables, loops, and branching.

2. Give as much time to words as to numbers, and give some time to music.

Far too many programming texts are devoted almost entirely to calculations using numeric variables. This is a surefire way to turn off many girls, minorities, and humanists. (That's a large group to lose all at once!)

After graphics, which should reach nearly everyone except people with sight problems, the manipulation of words and music will probably help to hook most of them for good. There is no reason that you need to teach about variables, loops, subroutines or procedures, or recursion using only mathematical concepts. These programming ideas are equally important in these other media as well, and the use of the concepts in more than one medium should give a boost to the student's understanding (Dickson, 1985).

3. Try to provide a context in which students need to know a command or concept to make their lives easier or make new things possible.

The moving graphic is an excellent assignment in this category. To make a graphic move you need to draw it, erase it, and then draw it in a new place. No one wants to type the lines of code for the object over and over again; loops and variables are grabbed up eagerly by students working on such a project. Students will eventually need to know something new in order to make their project do something neat. Kids begin to collect information and trade it; knowledge becomes important to them, something they seek on their own (Papert, 1980).

4. Accept both top-down and bottom-up programming styles.

Sometimes the preference for top-down programming is the teacher's attempt to make a scarce resource—the computer—go further by making the kids work the assignment out on paper first. That's certainly a realistic concern. But if at all possible, use sign-up sheets or some other tactic for maximizing hands-on computer time. Let students work at the computer and "negotiate" with it, trying a little of this, a little of that, until they see how it works and what they like (Turkle, 1984). They'll get immediate feedback, which we all know helps people learn more quickly than delayed feedback.

One factor that often prevents teachers from allowing bottom-up programming, even among beginners, is perceived pressure from the university. Although you may have heard college professors complain that pre-college teaching of programming is so sloppy that it makes teaching college computer science more difficult, the evidence is against that claim. In fact, beginning college computing courses tend to do better than those who have not (McGee et al., 1987). Furthermore, almost all college programming students start with an *introductory* computer science course, even if they've already had programming in high school. So unless you are teaching a special Advanced Placement course, it is not your job to prepare them for advanced courses in computer science. Colleges and universities take care of that. It is your job to prepare them for a beginning course in computer science, one they won't even take unless they understand and enjoy your course.

5. Let people work together on projects.

I know you have to grade them and you probably worry about some kids sitting back while others do all the work. There are a number of ways around this, including group grades, testing what they learned, and follow-up solo projects. At the very least find a way to let your students work on some things together as a group on a project of their own choosing (within guidelines). Girls, especially, have a difficult time getting involved with computing because it's such a solitary activity sitting there with a machine. When it becomes a social activity, girls enjoy it more (Sanders & Stone, 1986). Furthermore, you will be preparing students for the real world, since professional programmers (and indeed many other professionals) often work in teams.

6. Give students a framework for an assignment that allows lots of room for individuals decoration. Show outstanding products to the entire class. Make a disk of good projects for students to copy.

I have found this to work at all ages. When individual variations are allowed, students tend to learn more and work harder than when they all have to do the same thing. Decide what it is that's important for them to work on, and then give them an assignment that sets them free in that environment. For example, have students draw a picture using letters and other symbols on the keyboard. Or have them draw a flock of something—three rows, each containing a different size and number of some shape. Or something that combines picking a random number with the use of graphics or sound.

7. Study some neat programs and let students modify them.

This is basically a different approach to the same idea. Let students make their mark on a program that is famous or interesting or more difficult than what they could have written themselves, as for example the ELIZA program. Let them make this program as silly or as idiosyncratic as they like—they will still have to understand how it works in order to do that. Such programs can spark interesting class discussions. Furthermore, students will probably come to see how important it is for code to be readable and well documented.

8. Use motivating, interactive assignments like games and lessons.

Have students program their own version of a game, like Guess My Number, Hangperson, or Mastermind (some of these are tougher than others). It is a good opportunity to discuss a flowchart of events, discuss problems with certain pieces of code, and see problem solving in action. Again, students may jazz it up any way they like once they have the basic structure working. Another assignment that works is to program a drill that students can use for another subject. The fun and the interactivity of games and drills seem to get students involved at a higher level than programming calculations.

Conclusion

The National Science Foundation and many of the school reports give high priority to educating a citizenry to be literate in math, science, and computing. These reports place a special emphasis on the encouragement of girls and minorities in these fields. It is unlikely that girls and minorities will go into math, science and computing if we do not make an entry point that is interesting and attractive. But we can! Computing does not need to be taught in the limited fashion of the past. At the pre-college level the best thing we can do for our kids is to help them to understand and be interested in fields we would like them to be able to choose later on.

Diskette Available

Over the years I have collected and designed a lot of good assignments in BASIC, Logo, and Pascal. I have put these on a 5 1/4" for the Apple II series. For a copy, send a disk and return postage, and specify whether you would like the BASIC-Logo disk, the Apple Pascal disk, or both. If you request both, I will put the Pascal programs on the back.

Dr. Diane McGrath, Center for Science Education, 261 Bluemont Hall, Kansas State University, Manhattan, KS 66506; Bitnet: dmcgrath @ ksuvm.

References

Dickson, W. P. (1985, May). Thought-provoking software: Juxtaposing symbol systems. *Educational Researcher*, 14(5), 30-38.

McGee, L., Polychonopoulos, G. & Wilson, C. (1987). The influence of BASIC on performance in introductory computer science courses using Pascal. *SIGSCE Bulletin*, 19(3), 9-37.

Papert, J. S. (1980). *Mindstorms: Children, computers and powerful ideas.* New York: Basic Books.

Sanders, J. S. & Stone, A. (1986). *The neuter computer.* New York: Neal-Schuman.

Turkle, S. (1984). *The second self: Computers and the human spirit.* New York: Simon and Schuster.

CHAPTER 6

Solving Mathematical Problems in BASIC

BASIC is often been called "the language of microcomputers" because it is included at no extra cost, along with the computer's operating system, with most microcomputers sold. The language predates microcomputers, however. It was developed in the 1960s by John Kemeny and Thomas Kurtz at Dartmouth for mainframe computers as a nonthreatening way of introducing programming concepts to social science majors and as a tool for developing programs that would be useful in their professions, including simulations. In addition to the mathematical capabilities common to most programming languages, BASIC was designed to allow users to print text on a teletypewriter or screen and to allow easy evaluation of text inputted from the keyboard. These tasks were difficult to do with other languages available at the time—namely Fortran and Cobol—since up until that time little thought had been given to using computers for anything other than number crunching. Because of its text-handling capabilities, and, more recently, its graphics capabilities, BASIC was and continues to be a popular language for developing CAI.

When manufacturers such as Radio Shack, Commodore, Apple, and IBM were designing microcomputers to be sold to the public in the late 1970s, they needed an easy-to-use programming language to accompany their machines. BASIC was the natural choice because of its simplicity and ease of use. Early versions of the language were scaled down to fit into the limited memory space available in the microcomputers. Computer manufacturers, however, soon found that they needed to expand the language. For one thing, mainframe BASIC provided no way to create screen graphics. Since no graphics standards for BASIC had been set, each manufacturer devised its own methods and commands for creating screen graphics. Finally, in 1987, the American National Standards Institute (ANSI) established a new set of standards for BASIC that were based on IBM's version of the language and that included graphics standards and added many commands and techniques common to modern, structured languages, such as Pascal.

BASIC—especially the newest versions such as QBasic, Quick BASIC, Turbo BASIC, and True BASIC—is an excellent tool for exploring mathematics at the upper elementary, junior high, and high school levels. There are even possibilities for children in the middle elementary grades and lower, as discussed by Patricia S. Wilson in the article at the end of this chapter. BASIC contains the necessary commands and functions for doing a large variety of mathematical manipulations and the new versions of the language have the structured programming features that foster good programming habits and that set the stage for further study of computer programming and computer science.

This chapter introduces BASIC programming concepts with an emphasis on mathematical problem solving. The discussion focuses on and gives programming examples in the new BASICs, however, where differences between the new and older versions of BASIC; (AppleSoft and IBM BASIC) exist, they are discussed and program examples are given. The discussion in this chapter does not cover the nonlanguage features of the various packages (the editors, windows, debugging features, etc.) except in general terms. Consult your package's manuals or tutorial diskettes for help in using your BASIC package.

The PRINT Statement

One of the most useful and easiest things you can do in BASIC is to print something on the screen. The PRINT command tells the computer to print what follows in quotes on the screen.

New BASIC   Old BASIC

```
PRINT "Hello, how are you?"    10 PRINT "Hello, how are you?"
PRINT "I am fine!"             20 PRINT "I am fine!"
```

This example points out the most obvious difference between the old BASICs (AppleSoft BASIC and IBM BASIC) and the new BASIC: the old BASICs require line numbers, whereas the new BASICs do not (they are optional). BASIC programs are often numbered by 10s so that, later, lines may be inserted, if need be. In the new BASIC, program lines may be inserted using the program's editor, in a way similar to inserting lines with a word processor. In the old BASICs, if you wish to insert lines between previously written lines, simply type it at the bottom of the program with its line number. When you list the program, the line will be inserted in its proper place.

When using one of the new BASICs, you enter an environment similar to that of a modern word processor with pull-down menus and different windows for viewing and editing a program, for printing its output, and for listing program information, and so on. If you want to run your program, for example, you use a mouse to pull down the run menu, then select RUN. If you wish to edit your program, you select EDIT from a menu. You may then scroll through your program using scroll bars, you may cut and paste, or you may search and replace as you can with a word processor. See Fig. 6.1 for an example of what the screen looks like in the IBM version of Quick BASIC.

With the older versions of BASIC, you must use operating commands to do the tasks mentioned above. To run a program, for example, you type RUN rather than selecting RUN from a menu or pressing a function key. To look at a part of your program, you type LIST to see the whole program, or LIST 20-200 to see from lines 20 to 200. In the new BASIC, you use the mouse or arrow keys to scroll through your program.

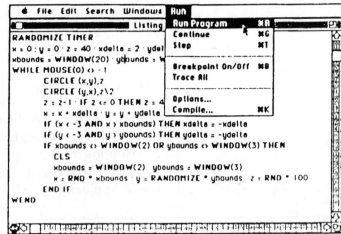

Fig. 6.1: Microsoft Quick BASIC screen

The PRINT statement (a statement consists of the command and the value it is acting upon—the complete BASIC *sentence*) may also be used to carry out arithmetic operations.

New BASIC	Old BASIC
`PRINT 3 + 5 * 2 ^ 4`	`10 PRINT 3 + 5 * 2 ^ 4`

In BASIC, as in most other programming languages, arithmetic expressions are evaluated in standard mathematical order—exponents are evaluated first, from left to right; then multiplications and division, from left to right; finally additions and subtractions, from left to right. Thus, in the examples above, the computer will print "83" as the output. The * symbol is used to indicate multiplication and the ^ to indicate exponent. Therefore, 2 is raised to the 4th power; the result, 16, is multiplied by 5 to obtain 80; then that result is added to 3.

To override the standard order of operations, use parentheses. The computer evaluates the expression inside parentheses first and if parentheses are inside parentheses, the innermost are evaluated first. Thus, the output to

```
PRINT 2.5 * (7 - 5)
```

will be 5.

Before we present a full-length (albeit, short) program, a couple of housekeeping commands: if you wish to start with an uncluttered, fresh screen and not be distracted by previous output or program listings, add a CLS (new or IBM BASIC) or HOME (AppleSoft) statement to your program. Finally, BASIC programs should end with the END statement. Note that computer programs are executed one line at a time, in the order in which they appear in the program, unless a statement in the program commands the computer to jump to another location.

The following is a BASIC program that prints "Hello, how are you?" then computes and prints the answer to 28 * 16 - 200 / 10 ^ (4 - 2):

New BASIC

Old BASIC

```
CLS                              10 CLS      (HOME on the Apple)
PRINT "Hello, how are you."      20 PRINT "Hello, how are you?"
PRINT 28*16-200/10^(4-2)         30 PRINT 28*16-200/10^(4-2)
END                              40 END
```

The sequence of characters between the quotation marks in the first PRINT statement is called a "string." The numbers and operations in the second PRINT statement is called an "expression."

After this program has been entered, you may execute it by selecting the RUN option from the appropriate menu if you are using a new BASIC, or by typing RUN (without a line number) after the end of the program if you are using Apple or IBM BASIC. If there is a syntax error in your program, the program will stop running, put the cursor in the line where the error occurred, and print an error message. In the new BASICs, you may use the delete key to get rid of the offending part of the line and/or insert the correct characters. In Apple or IBM BASIC (also called GW BASIC), simply retype (correctly) the incorrect line, including the line number. The new, corrected line will replace the old line in the program. The output from the program above should appear as in Fig. 6.2.

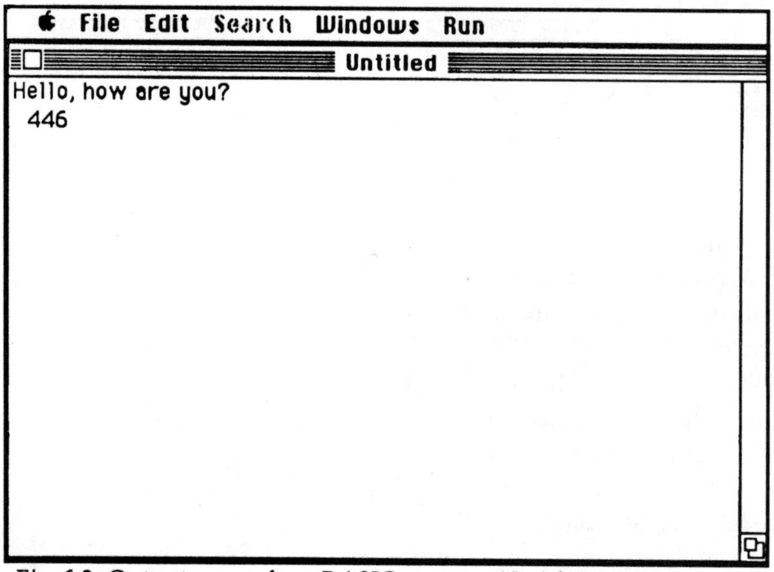

In the remainder of this chapter, unless there is a difference between the new and old versions of the program, only the new BASIC version will be presented. To convert the new BASIC programs listed below to AppleSoft or IBM BASIC, simply add line numbers at the beginning of every program line.

```
 File   Edit   Search   Windows   Run
                    Untitled
Hello, how are you?
 446
```

Fig. 6.2: Output screen from BASIC program (Quick BASIC, Macintosh Version)

If you wish to include more than one string or computation in a PRINT statement, separate them with semicolons:

```
    PRINT "The answer to 7 divided by 3 is ";7/3
or PRINT 27; " times "; 35 ; " equals "; 27*35
```

Note that in the second PRINT statement, the values, 27 and 35, rather than the strings, "27" and "35," are printed—an invisible difference here, but important elsewhere, including spreadsheets or data base managers. Also, note the spaces inside the quotation marks. Spaces inside of quotation marks are printed on the page. Outside of quotation marks, spaces are ignored. Thus,

```
    PRINT 27;    "times"   ;   35   ;    "equals";    27*35
```

would print the following on the screen

```
27times 35equals 945
```

(most versions of BASIC automatically leave spaces before and/or after all numbers) whereas

```
PRINT 27;" times  ";35;" equals  ";27*35
```

would print

```
27 times   35 equals   945
```

To position the output in precise locations on the screen or paper, use TAB() in the PRINT statement. The number in parentheses after TAB tells the computer where to begin typing the string or number following it.

```
PRINT TAB(10) "I like Math."
```

will print "I like Math," starting in column 10.

```
PRINT "A"; TAB(15) "B"; TAB(30) "C"
```

will print "A" at the left edge of the screen in column 1, "B" in column 15, and "C" in column 30. Most computers print 80 columns and 24 rows of text on the screen. Older Apples print only 40 columns on the screen. The following program uses TABs to print test scores and averages in columns.

```
PRINT "Name";Tab(10) "Test 1";Tab(20) "Test 2"; Tab (30) "Test ¹ 3"; Tab(40) "Average"
PRINT "John";Tab(10) 35;Tab(20) 40;Tab(30) 50;Tab(40) (35+40+50)/3
PRINT "Juanita";Tab(10) 44;Tab(20) 49;Tab(30) 47;Tab(40) (44+49+47)/3
PRINT "Jim";Tab(10) 26; Tab(20) 45;Tab(30) 39;Tab(40) (26+45+39)/3
END
```

Pupil Activity 6.1

Objective
Use the print statement to find answers to computational problems.
Grouping
One or two pupils.
Materials
Computer, BASIC.
Procedure
The students are given exercises or problems involving computations. They should use appropriate problem-solving strategies (e.g., restate the problem, plan a program to solve it, write the program, run and debug it, evaluate the results, make necessary program changes if the output is not correct).

---

¹ Do not press RETURN here. Keep typing until you have completed the PRINT statement.

Sample Problems

1. The population of California is 30,456,287. It is expected to grow by 10% in the next 5 years. What will be its population 5 years from now? Write a program that calculates and prints the answer in a sentence.

2. List the names and populations of the 10 most populous cities or towns in your state. Write a program that prints the names of these cities, their current population, their population in 5 years (assume that it will increase by 1/10 from their present population) and their population in 10 more years (assume a 1/5 increase), all in columns as in the program above (the one that prints data in columns using TABs). Print appropriate column headings at the top of each column.

Variables

The tasks done with the PRINT statement in the problems above really do not tap the power of the computer—they could have been done just as easily with a typewriter and a calculator. It is the use of *variables*—the temporary storage of computational results for use later in the execution of the program—that differentiates computers from calculators and typewriters and that gives them their power. In BASIC, values are assigned to variables in LET statements. The variable to which you are assigning a value (or the results of a more complex expression) is always by itself to the left of an equal sign. To the right is a value or expression.

```
LET SCORE = 25
LET NUMBER = 8 ^ 2
```

The first BASIC line assigns the value of 25 to the variable, SCORE, while the second assigns the results of the computation, 8 to the 2nd power, to NUMBER. Note that the LET command is optional in most versions of BASIC. These lines are used—without the LETs—in a short program below. The fourth line adds 1 to the current value of X, then stores the new value in X. The fifth line prints the values stored in the variables SCORE, NUMBER and X.

```
SCORE = 25
NUMBER = 8 ^ 2
X = 100
X = X + 1
PRINT SCORE; NUMBER; X
END
```

The following program calculates Joan's pay for a week at her job:

```
CLS
PAYRATE = 6.75
HOURS = 22.5
TAXES = .175
PRINT "Jane's takehome pay for the week of Jan. 18 was $";[2]
   PAYRATE * HOURS * (1 - TAXES)
END
```

---

[2] Do not press the RETURN or ENTER button here. Keep typing until you reach the end of the PRINT line. The PRINT line is divided here and the second part of this print line is indented from the left margin for clarity only.

Strings (sequences of characters that may not be used in computations) may also be assigned to variables. String variable names must end with a $.

```
CLS
LName$ = " Sanchez"
FName$ = "Jose"
Complement$ = " good student."
PRINT FName$;LName$;" is a";Compliment$
END
```

Variable names should be descriptive, but not so long that they are difficult to type. They must start with a letter, but after the first character, either letters or numbers may be used. Special characters such as *, (, %, and so on, should be avoided in variable names since many of them have special meanings in BASIC. Although most versions of BASIC allow variable names of 40 characters or more, AppleSoft BASIC only recognizes the first two characters of a variable name.

Pupil Activity 6.2

Objective
Use variables to solve problems.
Grouping
Individuals or pairs of pupils.
Materials
Computer, BASIC.
Procedure
Write programs to solve the following:
1. Find the volume of a box 45 centimeters long, 28 centimeters wide, and 36 centimeters deep. Assign the dimensions of the box to the variables, LENGTH, WIDTH, and DEPTH. Use the variables to calculate the volume and assign the results to the variable VOLUME. Print the results, followed by "cubic centimeters."
2. The Adams family is planning a 9376 mile trip in their RV this coming summer. Mr. Adams estimates that their RV gets 6 miles a gallon and that gasoline will average $1.25 per gallon during the summer. How much will they spend for gasoline on their trip? Use variables to solve this problem.

Counted Loops

Suppose you wanted to print "I love riding my mountain bike" 250 times on the screen. One way to do it would be to write a program such as the following:

```
PRINT "I love riding my mountain bike"
PRINT "I love riding my mountain bike"
PRINT "I love riding my mountain bike"
   Etc. for 247 more lines
```

Or, suppose you want to find out how much your house would be worth after 25 years if its value increases 10% each year and it is currently worth $125,000. One way would be to write a program such as the following:

```
Value = 125000
Value = value + value * 1.1
Value = value + value * 1.1
Value = value + value * 1.1
   Etc. for 22 more lines
```

Of course, you might as well use a typewriter for the first example or a calculator for the second if you are going to use this approach—writing a program saves no time or energy whatsoever! Fortunately, BASIC, like other programming languages, has structures that allow you to repeat a given set of commands a desired number of times. Old and new BASICS all contain the FOR-NEXT statement. This statement allows you to set up *counted* loops. In its simplest form, the FOR-NEXT loop starts counting from the initial value (1 in the example below), to the final value (250 in the example) by 1s.

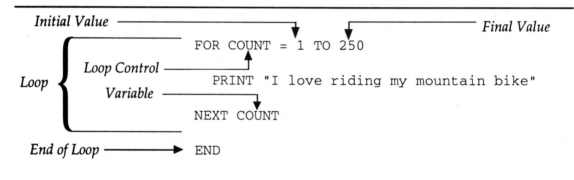

Figure 6.3 The FOR-NEXT loop.

And, each time it executes the body of the loop (the PRINT statement in this example) the print line is executed. The command FOR marks the beginning of the loop and NEXT marks the end. COUNT in this program is the *loop control variable* and can be any legal variable name. The loop control variable must appear immediately after the FOR and NEXT commands.

The loop control variable may be used within the body of the loop (it should not, however, be modified within the loop). For example, you might wish to print its value every time the loop is executed.

```
FOR COUNT = 1 TO 250
   PRINT "The number is ";COUNT
NEXT COUNT
END
```

FOR-NEXT loops may be used to count by different increments than 1 by adding STEP to the FOR line. Or, the initial value may be set to a number other than 1. They can even be used to count backwards. The following are some examples:

```
FOR X = 10 to 200 STEP 10
   PRINT X, X^2 ³
NEXT X
END
```

Step value—each time the loop is executed (after the first time) 10 is added to X.

Fig. 6.4 shows the output for this program.

```
FOR C2 = 10 TO 5 STEP -.25
   PRINT "Counting"
NEXT C2
END
```

This program will print "Counting" 21 times. The loop control variable is initially set to 10, then .25 is subtracted from it to make it 9.75, it then becomes 9.50, etc. This continues until C2 is less than 5). Note that if the initial number is larger than the final number, the STEP value must be negative.

Pupil Activity 6.3

Objective
Use counted loops to solve problems.
Grouping
Individuals or pairs of pupils.
Materials
Computers, BASIC.
Procedure
Write programs to solve the following:
1. Print the numbers from 100 to 10,000 by 100s.
2. Print the numbers from 280 down to 120 by 4s.
3. Solve Pupil Activity 5.3, #5 using BASIC.

10	100
20	400
30	900
40	1600
50	2500
60	3600
70	4900
80	6400
90	8100
100	10000
110	12100
120	14400
130	16900
140	19600
150	22500
160	25600
170	28900
180	32400
190	36100
200	40000

```
Press any key to continue
```

Fig. 6.4: Output for FOR-NEXT program

Keyboard Input

Suppose that you purchased a cookbook in Britain and wanted to convert all the recipes in it from metric units to American units. An appropriately written computer program would be ideal for this task. The most useful type of program to have would be one that would allow you to do any number of metric-American conversions by simply typing each metric unit from the keyboard as the program was running. The program would continue doing metric-American unit conversions for you until you asked it to stop. The INPUT command allows you to enter data from the keyboard into a program while it is running. (We will discuss later how to get the computer to repeat a task over and over.)

---

³ If a comma is used to separate elements in a PRINT statement, the next item after the comma is printed in the next print field on the screen. The print fields in the IBM begin in columns 1, 15, 30, 45, and 60.

When the computer encounters an INPUT command, it stops, prints a "?" on the screen, waits for something to be typed, than assigns this value to the variable following the INPUT command.

```
PRINT "Please type your first name."
INPUT NA$
PRINT "Hello, ";NA$;
PRINT "If you type the base and height of a triangle"
PRINT "and the kind of unit, I will tell you its area."
PRINT "Please type the length of its base.
INPUT BASE
PRINT "Please type its height."
INPUT HEIGHT
PRINT "What kind of unit?"
INPUT UNIT$
PRINT "The area of a triangle with a base of ";BASE
PRINT "and a height of ";HEIGHT;" is "
PRINT BASE * HEIGHT/2;" square ";UNIT$;"."
END
```

BASIC also allows you to put more than one variable after the INPUT command:

```
INPUT BASE, HEIGHT, UNIT$
```

Data can either be input in a line, separated by commas, or each value can be inputted separately followed by the RETURN or ENTER key.

Decision Statements

A decision statement carries out one of a number of alternative actions depending on the status of a condition. The simplest and most common kind of conditional is the IF statement. If the condition in the IF statement is true, the action(s) after the THEN are executed.

```
IF COUNT = 2 THEN PRINT "This is your second try."
```

Condition (can be evaluated as to whether it is true or false)

Fig. 6.5: The BASIC IF statement

If the condition (COUNT = 2) is not true, then the PRINT statement is not executed and the computer goes to the next line in the program.

An IF statement may also specify which action to execute if the condition is not true.

```
IF ANS$ <> "YES" THEN C5=C5+1: PRINT "Good, let's continue."
    ELSE C5 = 0: PRINT "Sorry." 4,5
```

In a single-line IF statement, multiple actions after the THEN or ELSE are separated by a colon. No colon is placed after the last statement prior to the ELSE.

```
IF X = 2 THEN Y = 5: PRINT "Try Another?" ELSE Y = 25:
    PRINT "That's all for today"
```

Single-line IF statements are hard to read and edit. Also, they are limited by the number of characters that can be typed in a line of program (around 250, depending on the computer). The new BASICs support multiple-line IF statements, a significant improvement over the single-line IF statements. They are easier to read and debug and may contain as many statements as needed. Multi-line IF statement must end with an END IF.

```
IF ANS$ <> "YES" THEN
  C5 = C5 + 1
    PRINT "Good, let's continue."
ELSE
  C5 = 0
    PRINT "Sorry."
END IF
```

IFs may be used to evaluate the value of variables to see if a particular action should be carried out.

Problem Example
Write a program that accepts two numbers from the keyboard and multiplies them if the first number is greater than 10.

Solution
```
PRINT "Type a number larger than 10."
INPUT A
PRINT "Type another number."
INPUT B
IF A > 10 THEN PRINT "The product of ";A;" and ";B;" is";A*B
END
```

Some of the possible conditions that can occur after the IF command are as follows:

```
IF COUNT > 6      (COUNT is greater than 6)
IF ANS$ <> "YES"  (ANS$ is not equal to "YES")
```

---

[4] The RETURN key is not pressed until the entire line is entered. In AppleSoft and IBM BASIC, one line number is used for the entire IF statement.
[5] In AppleSoft BASIC, the IF statement does not allow an ELSE.

```
IF X < 3                  (X is less than 3
IF POP3 <= 95             (POP 3 is less than or equal to 95)
IF X < 4 AND Y$ = "BOB"   (X is less than 4 and Y$ is equal to BOB)
IF P3 = 10 OR P4 < 16     (P3 equals 10 or P4 is less than 16)
```

Pupil Activity 6.4

Objective
Use keyboard input and decision statements to solve problems; find areas in square centimeters; calculate averages.
Grouping
Individuals or pairs of pupils.
Materials
Computer, BASIC, centimeter grid paper.
Procedure
Ask the students to solve problems like the following:
1. Write a program that gets 10 lengths (use a FOR-NEXT loop) in centimeters from the keyboard and converts them to inches (1 inch = 2.54 centimeters).
2. Write a program that gets a person's name and height in inches from the keyboard. The program uses one or more IF statements to state whether the person is "Tall" or "Not so tall."
3. Have five of your friends find the area of their right hand using centimeter grid paper (put your fingers together, place your hand on the grid paper, draw around the outside of your hand, then count the number of centimeter squares inside the enclosed region). Write a program that allows you to input the area of each hand, and then calculates the average area. Write another program that gets a person's name and the area of his or her hand, then uses IFs to state whether or not a person's hand size is "Average," "Bigger than average," or "Smaller than average."
4. Write a program that gets two numbers from the keyboard (e.g., 25 and 47). The computer then uses a FOR-NEXT loop to multiply the first number by the second number using repeated addition (e.g., adds forty-seven 25s together), and then prints the result.

Conditional Looping

Often times we wish to repeat a group of statements while a particular condition is in effect or until a particular condition is met. In the older versions of BASIC, we can use an IF statement to check to see if the condition is met, and then branch accordingly. With the newer BASICs, the DO LOOP provides such a capability. There are several forms of the DO LOOP:

```
DO WHILE Condition
   Statements
LOOP
```

In this form, the loop is executed while the condition is true. The LOOP command indicates the end of the loop. When the condition becomes false, the loop is terminated and the line after the LOOP command is executed. Note that the condition is checked before the loop is executed. If it is false, the loop is not executed at all. Here's an example:

```
X = 25
DO WHILE X < 50
      PRINT "The value of X is ";X
      X = X + 5
LOOP
PRINT "The loop is finished."
END
```

The second form of the DO LOOP is executed while the condition is false (i.e., until it becomes true).

```
DO UNTIL Condition
      Statements
LOOP
```

An example of this form follows:

```
INPT$ = "YES"
COUNT7 = 1
DO UNTIL COUNT7 >= 10 OR INPT$ = "NO"
   PRINT "Ten to the power of ";COUNT7;" is ";10^COUNT7
   COUNT7 = COUNT7 + 1
   PRINT "Would you like to see the next power?"
   INPUT INPT$
LOOP
END
```

The WHILE or UNTIL can also be placed at the end of the loop (some versions of BASIC allow a condition at both the beginning and at the end of a DO loop). That way, the condition check is done after the loop is completed rather than before it starts. The preceding program can be redone to make it slightly more efficient by placing the UNTIL at the end of the loop.

```
COUNT7 = 1
DO
   PRINT "Ten to the power of ";COUNT7;" is ";10^COUNT7
   COUNT7 = COUNT7 + 1
   PRINT "Would you like to see the next power?"
   INPUT INPT$
LOOP UNTIL COUNT7 >= 10 OR INPT$ = "NO"
END
```

If the condition is placed at the end of the loop, the loop is automatically executed at least once.

The following program example is a modification of the program on page 178. It gets the base and height of triangles and finds their areas until the user types "STOP." The IBM BASIC/Applesoft version of the program uses an IF-THEN at line 180 to set up a conditional loop.[6]

New BASIC Version

```
PRINT "Please type your first name."
INPUT NA$
PRINT "Hello, ";NA$;
PRINT "If you type the base and height of a triangle"
PRINT "and the kind of units used, I will tell you its area."
DO
   PRINT "Please type the length of the base of a triangle.
   INPUT BASE
   PRINT "Please type its height."
   INPUT HEIGHT
   PRINT "What kind of unit?"
   INPUT UNIT$
   PRINT "The area of a triangle with a base of ";BASE
   PRINT "and a height of ";HEIGHT;" is "
   PRINT BASE * HEIGHT/2;" square ";UNITS;"."
   PRINT "To quit, type 'STOP', to find the area of another
        triangle, press the ENTER key."
   INPUT ANS$
LOOP UNTIL ANS$ = "STOP"
END
```

Apple/IBM BASIC Version

```
10    PRINT "Please type your first name."
20    INPUT NA$
30    PRINT "Hello, ";NA$;
40    PRINT "If you type the base and height of a triangle"
50    PRINT "and the kind of units used, I will tell you its area."
60    PRINT "Please type the length of the base of a triangle.
70    INPUT BASE
80    PRINT "Please type its height."
90    INPUT HEIGHT
100   PRINT "What kind of unit?"
110   INPUT UNIT$
120   PRINT "The area of a triangle with a base of ";BASE
130   PRINT "and a height of ";HEIGHT;" is "
140   PRINT BASE * HEIGHT/2;" square ";UNITS;"."
150   PRINT "To quit, type 'STOP', to find the area of another
triangle, press the ENTER key."
```

---

[6] IBM BASIC has a WHILE-WEND loop that is essentially a DO loop using a WHILE condition. Consult your reference manual for instructions on using this statement.

```
160   INPUT ANS$
170   IF ANS$ <> "STOP" THEN 60
180   END
```

Pupil Activity 6.5

Objective
Solve problems involving conditional looping.
Grouping
Individuals or pairs of students.
Materials
Computer, BASIC.
Procedure
Have the students solve the following problems:
1. Rewrite the first program in problem 3 in Pupil Activity 6.4 above so that a conditional loop is used to get areas of students' hands until the program user types "STOP." *Note:* You will need to set up a variable in the program to count the number of areas inputted.
2. Rewrite the second program (which tells whether a person's hand is "AVERAGE," etc.) so that the program uses a conditional loop to keep asking the user to input the area of someone's hand until the user types "-99."
3. Write a program which uses a conditional loop to find the sum of all integers from 1 to 100. Print the sum.

Graphics (IBM and New BASICs)

IBM BASIC and the new BASICs (including their Macintosh and Apple IIgs versions) all use essentially the same system of graphics. The grid on which you draw and the colors available do depend, however, on the type of machine (IBM or Macintosh), monitor and video card, and level of resolution you select. On IBMs, resolutions vary from 320 by 200 programmable dots to 640 by 480 dots, or more. Any IBMs with a color monitors will allow you to use SCREEN 1 graphics, the lowest level of resolution (320 by 200), but with 16 background colors and 4 foreground colors available at a time. To switch into this graphics mode, use the statement

```
SCREEN 1
```

immediately before you start listing your graphics commands. (*Note:* To return to text mode, use SCREEN 0.) Your screen will appear as in Figure 6.6.

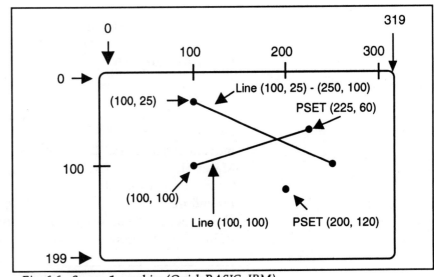

Fig. 6.6: Screen 1 graphics (Quick BASIC, IBM)

Once you are in the graphics mode, you may draw lines, points, boxes, filled boxes, circles or arcs. You may also fill any enclosed area with any available color. In this section, we will cover only lines, boxes, and circles.

To draw a point, use the PSET command. For example, the statement

```
PSET (200,120)
```

would draw a point 200 pixels[7] over from the left edge of the screen and 120 pixels down from the top. The PSET command is not very useful in most instances, especially in higher resolution environments—it is barely visible on the Macintosh, for example.
See Fig. 6.7.

To draw a line, you use (logically enough) the LINE command. Its format is

```
LINE (100,25)-(250,100)
```

Beginning coordinate   Ending coordinate

The line will be plotted in the default color or the last color used.

```
LINE (100,25)-(250,100),1
```
                              Color

Fig. 6.7: Format of LINE command

If you wish to draw a line from a previously plotted point (e.g., in another LINE command or using PSET) to another point, the first coordinate in the LINE statement may be deleted. The program lines

```
PSET (225,60)
LINE -(100,100)
```

would draw a line from (225,60) to (100,100). (See Fig. 6.6).

Pupil Activity 6.6

Objective
Use graphics coordinate system; draw geometric shapes.
Grouping
Individuals or pairs.
Materials
Computer and BASIC.
Procedure
Have your students use the LINE command to draw polygons—squares, right triangles, obtuse triangles, trapezoids, parallelograms, etc.

---

[7] A *pixel* is a programmable point on the screen.

To change the color of the background color, use the COLOR command.

```
COLOR 5
```

sets the background to color number 5. The following is a listing of the background colors available and their corresponding numbers:

```
0 Black    4 Red      8 Grey          12 Light red
1 Blue     5 Magenta  9 Light blue    13 Light magenta
2 Green    6 Brown    10 Light green  14 Yellow
3 Cyan     7 White    11 Light cyan   15 White
```

Table 6.1: Background colors

The foreground (drawing) colors available depends on which palette has been selected. Two palettes are available and are selected by adding a second parameter to the COLOR statement (see Fig 6.8).

```
COLOR 0,    1
      ↑          ↑
Background color  Palette
```

Fig. 6.8: The COLOR statement

Only two palettes are available in SCREEN 1 graphics. If Palette 0 has been selected, you may draw in green, red, brown, or the background color. If Palette 1 has been selected, you may draw in cyan, magenta, white, or the background color. See Table 6.2

```
Palette 0              Palette 1
0     Background       0     Background
1     Green            1     Cyan
2     Red              2     Magenta
3     Brown            3     White
```

Table 6.2: Foreground colors

Once a palette has been selected, a drawing color is selected by adding a number after a comma in a line, circle, or other graphic command. Fig. 6.9 shows a program for drawing a magenta line on a white background.

```
COLOR 7,1

LINE (100,100)-(150,150),2
```

Fig. 6.9: Program for drawing in color

The LINE command is also used to draw boxes (unfilled rectangles) and solid boxes. To draw a box, use LINE and the coordinates of the opposite corners of the box. (See Fig. 6.10).

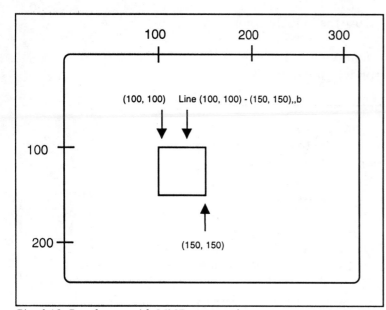

Fig. 6.10: *Box drawn with LINE command.*

```
LINE (100,100)-(150,150),,b
```

If you want to draw the box in a different color from the last color used, insert an appropriate number between the two commas in the statement.

```
LINE (100,100)-(150,150),2,b
```

Draws box with corners of (100,100) and (150,150)

To draw a solid box, add an f (for *fill*) after the "b."

```
LINE (100,100)-(150,150),,bf
```

The CIRCLE command allows you to position circles on the screen. Its format is shown in Figure 6.11.

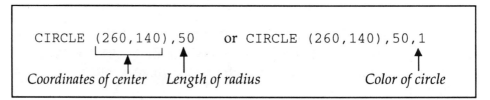

Fig. 6.11: *Format of Circle Command*

Note that the same command can be used to draw arcs and ovals. See your reference manual for details.

Unfortunately, there is no extension to the CIRCLE statement to fill it with a color. Instead, the PAINT statement will fill any enclosed shape. To fill a circle in red (assuming that palette 0 was in effect), use the following:

```
CIRCLE (260,140),50,2
PAINT (260,140),2
```

where the coordinate after PAINT is a point inside the shape which you wish to fill. Note that, unless you paint a shape in the same color in which it was drawn, the color may bleed out and fill a larger area.

Pupil Activity 6.7

Objective
Draw a scene with geometric shapes on coordinate system; use metric areas.
Grouping
Pairs or groups of three pupils.
Materials
Grid paper, Computers, BASIC.

Procedure

Have a computer math-art contest where groups of two or three pupils plan a graphics scene or abstract picture using regular geometric shapes. An added requirement could be that the total area of the shapes used be between 100 and 125 square centimeters on the screen. The students should plan and draw their scene on centimeter grid paper before beginning to write their program.

Subroutines

In programs that are only a few lines long, there is little difficulty in understanding or using them. Longer programs that are written as one big block of code, however, are difficult to read and debug. Programs are more readable when constructed in *modular* style (see the discussion of structured programming, below). One way to separate parts of the program that logically go together (e.g., a print routine, a set of statements that do a particular type of computation) is to place them in subroutines. A subroutine is simply a group of statements that is separated from the main program. A subroutine may be called from any part of the program, including from other subroutines. The command that calls a subroutine is the GOSUB. In the new BASICs, GOSUB command is followed by a *label*, a name that indicates the place where the subroutine begins:

```
GOSUB Calc
```

In the old BASICs, use a line number (the line number in front of the first line of the subroutine).

```
GOSUB 2000 8
```

Subroutines are placed after the END statement in the program. When the subroutine has been completed, the program returns to the next line after the GOSUB. (See Fig. 6.12).

```
PRINT "Please type the students test score."
INPUT T score
GOSUB Calc
GOSUB Prnt
END

Calc: ◄──────────────   The label Calc is followed
                        by a colon
   IF Tscore > 90 then
     Grade$ = "A"
   ELSEIF Tscore > 80 then
     Grade$ = "B"
   ELSEIF Tscore > 70 then
```

---

[8] Line numbers may also be used in the new BASICs, but there is no need to do so. In fact, they should be avoided

```
         Grade$ = "C"
   ELSEIF Tscore > 60 then
         Grade$ = "D"
   ELSE
         Grade$ = "F"
   END IF
RETURN
```

Prnt: ◄──────────── *The label Prnt is followed*
                    by a colon
```
   PRINT "The student's grade is ";Grade$
RETURN
```

Fig. 6.12 GOSUB examples

See the prime number program below (Program Example 2) for an example of the use of GOSUBs in Apple and IBM BASIC, and New BASIC.

REMARK Statements

When writing longer programs, or when writing a program over a period of time, it is often difficult to remember what certain parts of the program were designed to do. Also, when others (e.g., teachers, team members in programming projects) attempt to read your programs, it may be difficult for them to follow your code. REM (*remark*) statements were invented to allow programmers to put comments in programs for the purpose of explaining what certain blocks of the program were designed to do. A REM statement is one that the computer does not execute—it is in the program solely for documentary purposes:

```
   REM Programmer's name: John Q. Jones III
   REM Date:              October 22, 1991
```

New BASICs also allow an abbreviated form of the REM, the single quotation mark:

```
   ' This is a remark statement
   ' It will not be executed by the computer
```

See the discussion on structured programming below for further comments on the REM statement and programming examples containing them.

Structured Programming in BASIC

Imagine the house you would end up with if, without a clear idea as to the kind of house you wanted and without any architectural plans whatsoever, you purchased a stack of wood, nails, doors, windows, and such, and then immediately began poring cement, sawing and nailing things together. Chances are that you would be very unhappy with the resulting house—rooms, passageways, doors, closets, electrical outlets, and the like would be misplaced and unsightly. You would waste a great deal of time and materials and you would make many extra trips to lumber yards, hardware, and electrical stores to purchase materials as they were needed and as you redid parts of the house that were not properly built. Of course, building codes in most areas in developed countries do not allow this type of jerry-built construction. Also, the prospective homeowner does not want to risk his/her substantial investment with such lack of planning.

Writing a computer program, especially one that is long and complex, is a little like building a house. In order to end up with a functional and attractive end-product, you must first go through proper stages of design and then through the appropriate stages of construction. The art of writing clear, efficient programs that are relatively easy to understand (by both the programmer him/herself and others, such as members of a programming team, instructors, or supervisors) and to debug is called *structured programming*.

Structured programs have a number of features. They are as follows:

1. Top-Down. First of all, the major tasks of the program are outlined and directed from the main part of the program or the main procedure, then the details of the programs are written in *modules*—subroutines or procedures.

2. Modular. Rather than accomplishing the tasks of the program in a few, large blocks of code, the subtasks of the problem are developed and debugged separately in subroutines, procedures, or subprograms.

3. Internal Documentation. Structured programs have ample remark statements throughout to explain (1) the purpose of the program itself and the major subcomponents of the program, (2) the purpose of blocks of code (modules) within the program, (3) the list of variables used in the program and their purpose, (4) anything else in the program that might need clarification to the programmer him/herself or to anyone else attempting to decipher the code.

4. List Formatting. The programmer should use the features available in the language's editor to make the program more readable—the bodies of loops or blocks should be indented and aligned, each new statement should be placed on a new line rather than running the program together, and so on.

5. Structured Language Features. The programmer should make use of whatever structured language features are available and avoid unstructured features. For example, GOTOs should be avoided and, in their place, structured control statements such as DO-WHILE should be used.

6. Other Structured Programming Features. Different versions of BASIC have features that allow varying levels of structure. Some allow for long variable names—this permits the use of descriptive variable names, improving program readability and helping to avoid duplicating variable names. The new BASICs allow local variables in procedures or subprograms (not discussed in this chapter).

The following programs are examples of structured programs.

Program Example 1: Write a program to show that 1257 x 228 can be solved by adding 1257 two hundred and twenty-eight times.

New BASIC Solution

```
'Program to demonstrate that 1257 times 228 is the
'same as adding 228 1257s.
'
'******************** Variables Used ********************
'     COUNT — Controls DO loop
'     TOTAL — Keeps sum as 1257s are added
'**********************************************************

TOTAL = 0
COUNT = 0
DO
  TOTAL = TOTAL + 1257
  COUNT = COUNT + 1
LOOP UNTIL COUNT = 228

' ***** Print Results *****
PRINT "The sum of 228 1257s is ";TOTAL
PRINT "1257 times 228 equals "; 1257*228
END
```

Apple/IBM BASIC Solution:

```
10 REM Program to demonstrate that 1257 times 228 is the
20 REM same as adding 228 1257s.
30 REM
40 REM **************** Variables Used *****************
50 REM     COUNT — Controls the FOR loop
60 REM     TOTAL — Keeps sum as 1257s are added
70 REM ***********************************************
80 REM
90 TOTAL = 0
100 FOR COUNT = 1 TO 228
110    TOTAL = TOTAL + 1257
120 NEXT COUNT
130 REM ***** Print Results *****
140 PRINT "The sum of 228 1257s is ";TOTAL
150 PRINT "1257 times 228 equals ";1257*228
160 END
```

Program Example 2: Find 5 prime numbers larger than 2000.

New BASIC Solution:

```
' General Program to Determine if Entered Number Is Prime
'
' ***************** Variable List *********************
'     NUMBER — Input Amount
'     FLAG   — Set if Divisor Is Found
'     COUNT  — Loop Control
' ****************************************************

' ***** Main Program *****
PRINT "Type a number and the program will tell you if "
PRINT "it is prime or not.  To stop, type 0."
INPUT NUMBER
DO UNTIL NUMBER = 0
   FLAG = 0
   GOSUB calc
   GOSUB prnt
   PRINT "Type a number.  To quit, type 0."
   INPUT NUMBER
LOOP
END

' ***** SUB to Determine if number is Prime *****
calc:
   FOR COUNT = 2 TO SQR⁹(NUMBER)
     IF NUMBER/COUNT = INT¹⁰(NUMBER/COUNT) THEN
        FLAG = 1
     END IF
   NEXT COUNT
RETURN

' ***** SUB to Print Results *****
prnt:
   IF flag = 0 then
      PRINT number; " is a prime number."
   ELSE
      PRINT number; " is NOT a prime number."
   END IF
RETURN
```

---

⁹ SQR is a function that finds the square root of the number (or value stored in the variable) in parentheses. PRINT SQR(25) will find the square root of 25, then print it.

¹⁰ The INT function finds the integral value of the number or expression in parentheses. PRINT INT(32.456) will print 32. For more details (e.g., how INT handles negative numbers), check your reference manual.

Apple/IBM BASIC Solution

```
10 REM General program to determine if number is prime
20 REM
30 REM **************** Variables Used ****************
40 REM    NUMBER—Input amount
50 REM     FLAG — Set if a divisor is found
60 REM     COUNT —Loop control
70 REM *************************************************
80 REM
200 REM ************ Main Part of Program *************
210    PRINT "Type a number and the program will tell you
220    PRINT "if it is prime or not.  To quit, type 0."
230    INPUT NUMBER
240    IF NUMBER = 0 THEN 270
240    GOSUB 2000
250    GOSUB 3000
260 GOTO 200: REM ***** A necessary GOTO *****
270 END

2000 REM ***** SUB to determine if number is prime *****
2010 FLAG = 0
2020 IF NUMBER = 0 THEN 2060
2030 FOR COUNT = 2 TO SQR(NUMBER)
2040    IF NUMBER/COUNT = INT(NUMBER/COUNT) THEN FLAG = 1:
           GOTO 2060
2050 NEXT COUNT
2060 RETURN

3000 REM ***** SUB to Print Results *****
3010 IF FLAG = 1 THEN PRINT NUMBER " is not prime."
3020 IF FLAG = 0 THEN PRINT NUMBER " is prime."
3030 RETURN
```

The amount of effort involved in creating structured programs (indeed, the *degree* to which you can structure them) depends, to some extent, on the version of the language in which you are programming. Older versions of BASIC (e.g., early mainframe BASICs and AppleSoft BASIC) were developed before current theories and research findings about structured programming appeared. Because of its lack of structured features, it is virtually impossible to write a long, complex program in AppleSoft without GOTOs—modern, structured control statements such as DO WHILE and the structured IF-THEN ELSE are not a part of the language. In the early Logos, poor editors made it impossible to split a REPEAT statement into several lines, resulting in code that was often difficult to decipher. In AppleSoft and IBM-BASIC, the same thing was true with the IF-THEN statement. Also, in

AppleSoft BASIC, spaces were not preserved or were arbitrarily inserted into the program listing, making it impossible to indent and align blocks of statements.

New versions of BASIC such as QBASIC, Turbo, and True BASIC have eliminated these problems. They include a variety of new, structured commands such as DO WHILE, multi-line IF-THEN-ELSE, SELECT CASE, and subprograms to which parameters can be sent. They also have sophisticated editors, advanced debugging tools, windowing environments, online help, program outliners, and the like to greatly enhance the process of structured programming. New versions of Logo also have more sophisticated editors, use richer vocabularies, and allow multi-line statements.

Many K-8 teachers who have their students do programming/problem-solving activities in their classes may consider that teaching and insisting on structured programming strategies in their classes is a waste of time because the beginning programs children write are short and, often, easy to decipher. Also, children often rebel at the idea of writing programs in, what to them, appears to be an artificial format—they would rather begin writing code immediately starting with the first ideas they have; then, if their first impulse does not work, go back and try to patch up their first attempt. Teachers should be insistent—after all, we insist that our students follow the "correct" procedure when they do mathematical computations (e.g., in multidigit multiplication), even though they might be able to bypass many steps with easier problems. We want children to develop proper habits so that they will not run into difficulty later on when they encounter more complex problems. Likewise, the best way to become a good golfer is not to simply start playing, but to begin with lessons so that good form can be developed—it is difficult to break bad habits after they have been used for a long time.

Perhaps the most important thing you can do to encourage a structured approach to programming in BASIC is to use a modern, structured version of the language such as Quick, Turbo, or True BASIC. And, if you can use one of these in your class, insist on the use of its structured features and avoid the older, unstructured features (these new languages have preserved the older features of BASIC such as line numbers, GOTOs, the single-line IF-THENs, etc. in order to maintain compatibility with the older BASICs). If you *must* use an older BASIC such as AppleSoft, you will have to work harder to encourage structured programming habits in your students.

For those using older BASICs, follow the general structured programming guidelines listed on the previous pages—avoid GOTOs as much as possible, create modules using subroutines, make extensive use of REM statements to document your programs, use meaningful variable names, and format the listing of the program by leaving blank lines between program blocks (you may need to use empty REM lines for this) and indenting the body of loops and other blocks of code (in AppleSoft, colons can be used to indent, as shown in Fig. 6.13).

```
500 FOR C1OUNT = 1 TO 25
510 ::GOSUB 5000
520 ::GOSUB 7000
530 ::TOTAL = TOTAL + 1
540 ::FOR C2OUNT = 1 TO 10
550 ::::PRINT
560 ::NEXT C2OUNT
570 NEXT C1OUNT
```

Fig. 6.13 Indenting in AppleSoft BASIC

Programming languages provide excellent tools to solve mathematical problems. However, when using them, teachers must be careful to avoid the tendency to focus solely on the mathematical goal of the lesson at the expense of developing good programming/problem-solving habits. The efforts in developing structured programming habits will pay off in better organized, more readable programs. And, when following structured programming habits, students will be more successful in finding solutions to problems. Finally, structured programming corresponds closely to the mathematical problem-solving strategies discussed in Chapter 1 of this book—if a student goes through the proper problem-solving/structured programming procedures when solving a problem using a computer, chances are improved that he/she will use a more deliberate, organized approach when solving problems in a noncomputer setting.

There are other considerations when teaching programming in the classroom: How much should you tell the students and how much should you let them discover for themselves? How much cooperation should you allow between students when developing algorithms to solve problems? William Doyle, in his article at the end of this chapter, makes a case for teaching BASIC through discovery. Kenneth Kwajewski, in his article, advocates that instructors do a minimum of lecturing and a maximum of programming in BASIC programming classes.

Bibliography

Hergert, D. 1989. Microsoft Quick BASIC. Redmond, Wash.: Microsoft Press.

Presley, B. 1985. A Guide to Programming the IBM Personal Computers, Albany, N.Y.: Lawrenceville Press.

Presley, B. 1985. A Guide to Programming in AppleSoft. Albany, N.Y.: Lawrenceville Press.

Wiebe, J. H. 1981. BASIC programming for gifted elementary students. *Arithmetic Teacher* 28(7, March): 42-44.

Yarmish, R. and Yarmish, J. 1988. ANSI Structured BASIC. *SRA MIS Newsletter*. 1(2, Spring).

Discussion Questions

1. There are those who say that programming should not be taught at all in K-12 schools. Their rationale is that few people will ever become professional computer programmers. They say that we should, instead, teach children to use word processors and other computer tools that will be more useful to them in later life. How do you respond to this statement?

2. In what ways can the PRINT statement be used in the mathematics classroom?

3. What concepts used in programming languages are also used in other computer applications such as spreadsheets and data base managers?

4. What are the disadvantages of the old, unstructured IF-THEN statement when compared to the new, structured IF-THEN?

Activities

Activity 6.1

Objective
Write BASIC programs to solve problems in mathematics.
Materials
Computer, BASIC, printer.
Advance Preparation
Read appropriate sections of this chapter.
Procedure
Read this chapter carefully and do the suggested pupil activities throughout the chapter.

Activity 6.2

Objective
Develop an activity to help students understand a graphing coordinate system.
Materials
Paper and pencil.
Advance Preparation
Read the section on graphics in this chapter and Pupil Activity 2.1.
Procedure
Develop an activity similar to "Battleship" to help students understand the coordinate graphing system in BASIC.

Activity 6.3
Objective
Write a program that gives pupils practice with naming polygons.
Materials
Computer, BASIC.
Advance Preparation
Read this chapter and do the suggested programming activities.

Procedure

Write a BASIC program that describes a geometric shape and puts a picture of it on the screen. The program should then ask the student to name the shape. If the inputted answer is correct, print "CORRECT." If the answer is incorrect, give the student a hint (e.g., "starts with TRI"). If the student gets the answer wrong a second time, tell him/her the answer. Repeat for four other shapes.

Microcomputer Use in the Elementary School

by Patricia S. Wilson

Current software in elementary mathematics offers tutorial and practice programs as well as more open ended activities, such as simulations, games, utilities, and problem solving. In addition to software, elementary school students are using student- and teacher-made programs (BASIC or Logo) to explore mathematics. Owing to the wide range of lessons, researchers have found it necessary to adopt a variety of methods for studying mathematics learning, including case studies, classroom observations, and large-scale data collection. Although it is difficult to compare such diverse studies, results support the following conclusions.

• Young children can use software and can program in BASIC and Logo.

Primary students use the microcomputer in ways similar to a learning center. They can use and modify simple programs that involve graphics or familiar counting ideas, but they do not have the formal reasoning to deal with such abstract ideas as recursion or procedure.

Open-ended software and programming can help students develop problem-solving abilities.

Many of the gains were demonstrated in studies involving geometry or Logo graphics. Logo encourages rulemaking and self-direction that can transfer to noncomputer contexts. In addition, computer graphics help furnish a link from concrete to abstract ideas. Open-ended software and programming tasks encourage language experiences that increase students' creativity and help them to reflect on their own thinking.

• Mathematics instruction supplemented by drill-and-practice programs may improve mathematical skills .

Slightly more than half the recent studies showed improved skills using the microcomputer in addition to traditional classroom lessons. The remaining studies showed that microcomputer practice produced achievement equivalent to using traditional practice. Some evidence indicated that students who used exciting and colorful programs were dissatisfied when asked to return to paper-and-pencil practice. Although no studies showed a detrimental effect, results suggest that practice may not be the most efficient use of a limited supply of computers.

• Students using microcomputers have positive attitudes toward mathematics.

Both practice software and problem-solving activities contributed to an improvement in attitudes toward mathematics. Studies also reported positives attitudes toward computers.

• Children work together effectively on mathematics tasks.

Working in groups of two or three students, students instruct each other, negotiate, and collaborate. Students gain confidence and self-esteem while becoming less dependent on the teacher. Computer activities increase productive discussion and interaction among students.

• Students work on computer activities longer than other activities and talk more about the task.

Direct-answer questions produced talk about being correct or winning, whereas open-ended questions elicited wondering and hypothesizing. The increased discussion may contribute to increased success in solving problems.

• Using the microcomputer to teach mathematics is enhanced by teacher direction.

Research has been unable to measure the mathematics learning that may take place during unguided exploration on the microcomputer. Some teacher guidance enhances learning, but the nature and extent of the guidance is not yet clear.

In summary, researchers have been unable to detect extensive improvement in mathematical skills, but some evidence suggests that students acquire attributes related to success in mathematics. Students sustain attention on the task, use problem-solving strategies, gain self-confidence, and develop a positive attitude toward mathematics.

The following programs in BASIC simulate common growth patterns.

ADD/GROW calculates a constant growth each month by adding a specified amount. This pattern of growth could simulate (1) adding $2 to your piggy bank each month, (2) adding two new students to the school each month, or (3) growing 2 cm each month.

```
10    REM ADD/GROW
20    INPUT " MONTHS?" ;M
30    INPUT "GROWTH?" ;G
40    INPUT "BEGIN?";A
100   FOR X = 1 TO M
150   A = A + G
160   PRINT A
200   NEXT X
400   END
```

```
40    INPUT "BEGIN?";A
90    PRINT "AMOUNT",
         "GROWTH"
100   FOR X = 1 TO M
150   A = A + G
160   PRINT A,G
170   G = G*.5
200   NEXT X
400   END
```

```
MONTHS?5
GROWTH ?2
BEGIN?1
3
5
7
9
11
```

```
MONTHS?5
GROWTH ?2
BEGIN?10
12
14
16
18
20
```

```
MONTHS?5
GROWTH?2
BEGIN?1
AMOUNT  GROWTH
3        2
4        1
4.5      .5
4.75     .25
4.875    .125
```

```
MONTHS?5
GROWTH?2
BEGIN?10
AMOUNT  GROWTH
12       2
13       1
13.5     .5
13.75    .25
13.875   .125
```

MULT/GROW calculates an increasing growth each month, such as doubling or tripling the previous amount. (Modify ADD/GROW by changing line 150.) This pattern of growth could simulate (1) doubling the number of insects each month, (2) folding a piece of paper in half repeatedly, or (3) running twice as far each month.

```
10    REM MULT/GROW
20    INPUT "MONTHS?" ;M
30    INPUT "GROWTH?" ;G
40    INPUT "BEGIN?" ;A
100   FOR X = 1 TO M 150
150   A = A*G
160   PRINT A
200   NEXT X
400   END
```

```
MONTHS?5
GROWTH?2
BEGIN?1
2
4
8
16
30
```

```
MONTHS?5
GROWTH?2
BEGIN?10
20
40
80
160
320
```

By exploring each program and then by comparing programs, students can create their own theories about different kinds of growth and the mathematics that describes the growth.

The purpose of the activity is to encourage students to test their own theories by making the microcomputer work for them. The teacher can help students focus on the mathematical ideas by asking questions or offering suggestions. The following are examples:

1. Play with the program until you figure out what it is calculating.

2. Try to estimate the growth before the computer calculates it.

3. Draw a graph showing the growth each month.

4. Describe examples that could be simulated by each growth pattern.

5. What happens if you start with zero as the beginning amount?

6. Change the input to show how the number of months, the growth, and the beginning amount influence the final amount.

7. Compare the programs and figure out how they differ.

8. Modify the programs to develop a new pattern of growth.

REDUCED/GROW calculates the reduced growth by adding only half the growth of the previous month. (Modify ADD/GROW by adding lines 10, 90, 160, and 170.) This pattern could simulate roughly (1) plant growth, (2) growth of a person, or (3) any tapered growth.

```
10    REM  REDUCED/GROW
20    INPUT "MONTHS?";M
30    INPUT "GROWTH?";G
```

Organize students into groups of two or three so they can discuss their ideas. Provide paper and pencils for making tables and graph paper for making graphs. Allow sufficient time to generate ideas and to explore them. Offer suggested input to get students started. Suggest specific problems for investigation. Some suggestions follow:

1. How many times would you have to fold a piece of paper in half in order to have 256 thicknesses? Will you ever have exactly 400 thicknesses?

2. If your school grew by three people each month, when would you double your population?

3. If each year a twelve-foot tree grows half as much as it grew the year before, how much will it grow in ten years? What else do you need to know?

Bibliography

Battista, Michael T., and Kathleen J. Steele. "The Effect of Computer-assisted and Computer Programming Instruction on the Computer Literacy of High-Ability Fifth-Grade Students." *School Science and Mathematics* 84 (December 1984):649-58.

Clements, Douglas H. "Research on Logo in Education. Is the Turtle Slow but Steady or Not Even in the Race?" *Computers in the Schools* 2 (Summer/Fall 1985):63-71.

———"Effects of Computer Programming on Young Children's Cognition." *Journal of Educational Psychology* 76 (December 1984): 1051-58.

Clements, Douglas H., and Bonnie K. Nastasi. "Effects of Computer Environments on Social-Emotional Development: Logo and Computer-assisted Instruction." *Computers in the Schools* 2 (Summer/Fall 1985) 11-31.

Fuson, Karen C., and Kathleen T. Brinko. "The Comparative Effectiveness of Microcomputers and Flash Cards in the Drill and Practice of Basic Mathematics Facts." *Journal for Research in Mathematics Education* 16 (May 1985):225-32.

Michayluk, J. 0. "Logo: More Than a Decade Later." *British Journal of Educational Technology* 17 (January 1986):35-41.

A Discovery Approach to Teaching Programming

by William H. Doyle

This journal has recently featured articles that stressed the importance of having elementary school teachers and students learn to write simple computer programs. This emphasis suggests the need to share successful strategies for helping teachers to gain elementary programming skills. Teachers will then be more likely to introduce or offer direction for students' programming in their classrooms. One successful approach uses discovery learning and leads naturally to programming without creating inhibiting anxiety. The objective is not to produce expert computer programmers but rather to help teachers understand the elementary constructs of BASIC and to begin writing simple programs. The approach uses short programs that illustrate educationally sound uses of computers in the classroom. The central, elementary constructs of BASIC are taught through the simultaneous study of program listings on paper and program output on the monitor.

Description of the approach

I have used this approach as part of a ten-session (three hours each) course on using the microcomputer in elementary school mathematics instruction. Four sessions have been required to study the following BASIC constructs: FOR-NEXT, PRINT, GOTO, assignment, INPUT, IF-THEN, READ-DATA, variables (numeric and string), and graphics. Arrays and subroutines are individually presented to those teachers who desire to write programs that require such advanced concepts. Commands such as NEW, LIST, CATALOG, LOAD, and SAVE and skills such as editing and debugging are also introduced.

In session one the constructs FOR-NEXT, PRINT, and INPUT and variables are presented. Session two includes an introduction of the constructs GOTO, IF-THEN, and READ-DATA, In session three additional programs are run and studied to reinforce the concepts presented in the first two sessions. Projects are also assigned. The teachers are encouraged to identify concepts and skills that prove difficult for their students and to write short programs that will help teach those ideas. Graphics are presented in session four. In particular. the commands and techniques for plotting points, drawing lines, and creating motion are studied, extended, and applied.

The remaining six sessions are devoted to topics other than programming. Problem solving, software evaluation, curricular implications of the use of microcomputers, strategies for implementing the use of microcomputers, and an introduction to Logo are sample topics. However, time is regularly made available for consultation on programming problems, and occasionally a programming skill or construct is discussed with the group as a whole.

Apple II+ and IIe microcomputer systems (with disk drives) and Commodore PET systems (with tape drives) have been used most often. Two teachers at each machine is the optimal arrangement, partly because anxious teachers working in pairs support each other and challenge each other in the discovery process. However, because of limited resources, the course has, on occasion, been taught with three teachers at each machine.

Although the programs used are machine independent, operating-system commands (such as those for loading and saving programs) and graphics capabilities are machine dependent. I recommend that only one type of machine be used in the first four sessions, if possible. Lack of uniformity increases the potential for initial errors (usually associated with the operating system) that can reinforce any negative attitudes the teachers may have about computers.

The general procedure for the three sessions (first, second, and fourth) in which new ideas are introduced is as follows:

1. Load a program from disk or tape into the computer's memory and then run it.
2. Study the output on the screen along with a hardcopy listing of the program.
3. Encourage teachers to discover the causal relationship between BASIC constructs in the program listing and the output.
4. List conclusions on an overhead transparency.
5. Repeat steps 1-5 at least twice (using different computer programs) for each group of BASIC constructs taught.

Two additional strategies are frequently used during these lessons:

1. Teachers are asked to write programs to accomplish tasks closely related to those performed by the sample programs.
2. Teachers are asked to write programs to acomplish tasks closely related to those performed by the sample program.

Both of these teaching moves flow freely from the spirit of a discovery approach.

A Lesson Scenario

At this point, some sample programs and a scenario of what happens during a teaching session should be helpful. Figure 1 presents a listing and sample run for a version of the first program often used in the session.

Having loaded this short program, teachers type RUN with the caution of an infant taking his or her first steps. The results are invariably stimulating. As the machine "counts" to 10 and prints the square of each interger, teachers become fascinated by the scrolling display, even if they are uncertain as to its meaning.

A brief comparative study of the screen content and the program listing on paper leads to a barrage of good questions. The early evidence of anxiety disappears as teachers begin to discover the causal relationship between the simple BASIC commands and the display created by the mysterious, clearly powerful machine. The teachers have had a quick but gratifying encounter with a computer. They have commanded it to respond and have understood the constructs of the new language, BASIC.

A second example relates any who are still anxious about their confrontation with the machine. Figure 2 presents a program that was once used by an ingenious elementary school student to complete a disciplinary assignment.

After their amazement subsides, teachers give serious thought to an extension of the program: "How could this student instruct the computer to number the sentences?" Line 20 in our first example offers a substantial clue. Although the **exact** syntax is unavailable to them at this point, several teachers immediately see the need to have the value of N printed before the sentence proper. Figure 3 shows the line that effects the desired change.

Figure 2 An ingenious application

```
10 FOR N = 1 TO 100
20   PRINT "I WILL NOT
     TALK IN MR. JONES'
     CLASS"
30 NEXT N
40 END
```

Figure 3 A modification

```
20 PRINT N; "I WILL NOT
   TALK IN MR. JONES'
   CLASS "
```

It is time to challenge the teachers to write their first program. They are instructed to write a short program, using FOR-NEXT and PRINT that will make the machine count to 20 and display the results. Although the desired program will closely resemble the program in figure 1, the teachers have not considered the assignment to be too easy. The class typically breaks into subgroups, and within minutes, programs are written, entered, and tested. Spontaneously, some groups modify their programs to count to 1,000 or 1,000,000. (I am nearly always asked to help stop the machines that are taking forever to count to 1,000,000.) For those who have some difficulty in writing the program, hints given personally prove quite effective. The introduction of the STEP construct allows a further extension of the counting concept. Another flurry of activity follows as different counting multiples are tried spontaneously. Counting backward, counting by tenths, and so on, are further extensions. By this time most teachers are "hooked". The enthusiasm displayed by many seems to be contagious.

The other BASIC constructs are taught to the same rhythm, which stresses active involvement of the teachers. Teachers do more than read about INPUT; they provide input interactively, both for the computer and for class discussion.

I have found another teaching move helpful in solidifying what is done in class. Students should be encouraged to take program listings home, key in selected programs, run them, and then restudy the relationships between the listings and the printouts. This activity has proved helpful to many graduates of this course.

Figure 1 A first program

```
LIST

10   FOR N = 1 TO 10
20   PRINT N,N * N,3.5 * N
30   NEXT N
40   END

1    1      3.5
2    4      7
3    9      10.5
4    16     14
5    25     17.5
6    36     21
7    49     24.5
8    64     28
9    81     31.5
10   100    35
```

Examples

Figures 4-7 present four additional short programs that I have used. The subject matter identifies the authors' training. But experts in other areas of the curriculum can use programs illustrating other subjects.

The programs used should be "good" programs. They should clearly show the desired BASIC constructs. They should also illustrate instructionally valid applications (such as conceptual learning, problem solving, simulations, and games) of microcomputers in the classroom.

Figures 8-10 present three programs written by graduates of the course. None of these teachers had had prior programming experience.

Figure 8 presents a drill program for first and second graders. Figure 9 presents an example of a conceptual learning program for fifth and sixth graders. If a student can write a similar program to give the lowest common denominator, then that student understands the concept. Figure 10 presents a problem-solving program for gifted students in fifth and sixth grades. The problem is to list all combinations of the three symbols in a language used on another planet. The program is a small component of a creative, much longer problem-solving activity designed to stretch the student's ability to work with the concept of combinations. The concluding problem asks the student to find the number of license plates that can be created from six characters, three letters followed by three integers.

Much more sophisticated programs have been submitted, but the ones shown illustrate the average level of programming skill demonstrated. Most of the programs can be improved both in terms of programming style (efficiency, documentation, structure, etc.) and software design (screen formatting, freedom from external help, degree of student control, etc.). But the teachers have successfully completed a critical first step. Those who are interested can take additional courses in programming and software design. Although the majority of teachers do not have time to write software for their classes, they have a much better feel for how a computer works, are better prepared to help their students learn elementary programming, and are able to write programs specifically for the special needs of their students.

Figure 4 Simulation: Population Growth

This program simulates, on the basis of constant, current growth rates, when the population of Mexico will exceed the population of the United States.

```
5     REM POPULATION SIMULATION
10    LET US = 240000000
20    LET MEX = 68000000
30    LET YR = 1983
40    LET US = 1.009 * US
50    LET MEX = 1.028 * MEX
60    LET YR = YR + 1
70    IF MEX < US THEN 40
80    PRINT " POP US = ";US
90    PRINT " POP MEX = ": MEX
100   PRINT "IN YEAR ";YR
110   END

POP US = 441377864
POP MEX = 444672946
IN YEAR 2051
```

Figure 5 Concept learning: Limits

```
10  LET M = 10
20  FOR L = 1 TO 6
30    PRINT "1/" ;      M,1/M
40    LET M = M*10
50  NEXT L
60  END

1/10              .1
1/100             .01
1/1000            1E-03
1/10000           1E-04
1/100000          1E-05
1/1000000         1E-06
```

Strengths

Evaluations have shown that this approach has been successful in introducing teachers to the fundamental constructs of BASIC. Several reasons for this success can be identified. First, this method motivates teachers. It helps them bring their anxiety under control quickly. It allows them to use the machines immediately for a stimulating initial encounter that has a high probability of success. The first encounter is certain to go well, partly because loading, rather than typing, programs into the machine minimizes the potential for initial frustrating mistakes, such as syntax errors. Although the command to load a program must be typed, this one-line task has not proved to be difficult. So success is nearly automatic and motivation is strong.

Figure 6 Game:Integer search

```
10 PRINT " FIND AN INTEGER
      BETWEEN 1 AND 20 "
20 PRINT
30 PRINT " TYPE IN YOUR
      RESPONSE. "
35 PRINT
40 INPUT RE
50 IF RE < 5 THEN PRINT
     "TOO SMALL; TRY
      AGAIN ": GOTO 35
60 IF RE > 5 THEN PRINT
     " TOO LARGE, TRY
      AGAIN " : GOTO 35
70 PRINT
80 PRINT " CORRECT! "
90 END


FIND AN INTEGER BETWEEN
     1 AND 20
TYPE IN YOUR RESPONSE.

?10
TOO LARGE; TRY AGAIN
?3
TOO SMALL; TRY AGAIN
?5
CORRECT!
```

Figure 7 Problem solving: Probability

Find the probability of getting five heads in ten flips of a coin. Use the computer to simulte 100 repititions of the process of flipping a fair coin ten times. Then count the number of simulations that yield exactly five heads. This process is an example of the Monte Carlo technique.

```
10 FOR N = 1 TO 10
20   LET Y = RND(1)
30        IF Y <.5 THEN 60
40   PRINT "T" ;
50   GOTO 70
60   PRINT "H"
70 NEXT  N
80 END


THHTTTTTTT
HTTTTHHTHH
```

Second, the use of short programs does not overwhelm beginners. Teachers recognize that they can indeed write similar sets of instructions. But of even greater importance, they recognize the educational value of good programs and realize that they, subject matter experts who are in daily contact with their students, have the curricular expertise that is fundamental to quality education software.

A third strength of this discovery approach is that it gently nudges teachers toward writing their own programs. An appropriate heuristic for this task is to solve a simple problem first. As in the scenario presented earlier, teachers are first asked to modify an existing program. They are then ready to accept the more difficult challenge of writing a program as a natural extension of their work.

Figure 8 Drill and practice: Homophones

```
5  REM DRILL AND PRACTICE—HOMOPHONES
10    PRINT : PRINT "WORDS THAT SOUND ALIKE
20    FOR N = 1 TO 10
30    READ A$, B$, C$
40    PRINT : PRINT " TYPE A WORD THAT
         SOUNDS LIKE "
50    PRINT : HTAB (15) : PRINT A$
60    PRINT " INPUT I$
70    IF I$ = B$ OR I$ = C$ THEN 90
80    PRINT : PRINT "NO, TRY AGAIN": GOTO 60
90    PRINT : PRINT "THAT IS CORRECT!"
95    NEXT N
100   PRINT : PRINT "THAT IS ALL.  RETURN
         TO  YOUR SEAT."
110   DATA  SEE,SEA, SEA
120   DATA  BEAT,BEET,BEET
130   DATA  CENT, SENT, SCENT
140   DATA BLEW, BLUE, BLUE
150   DATA MEET,MEAT,MEAT
160   DATA WEIGH,WAY, WHEY
170   DATA PAIR,PEAR, PARE
180   DATA SUM,SOME,SOME
190   DATA SEW, SO,SO
200   DATA TWO,TO,TOO
300   END

WORDS THAT SOUND ALIKE

TYPE A WORD THAT SOUNDS LIKE
                        SEE
?SEA
THAT IS CORRECT!

TYPE A WORD THAT SOUNDS LIKE
                        BEAT
?BET
NO, TRY AGAIN
?BEET?
THAT IS CORRECT!
TYPE A WORD THAT SOUNDS LIKE
                        CENT
```

Figure 9 Concept Learning: Least common denominators

```
5       REM   LEAST COMMON DENOMINATORS
10      PRINT : PRINT : PRINT
30      PRINT "THIS PROGRAM WILL HELP YOU
        FIND"
32      PRINT "THE LEAST COMMON
        DENOMINATORS."
34      PRINT " OF ANY TWO DENOMINATORS."
35      PRINT
40      PRINT " TYPE IN TWO DENOMINATORS "
45      PRINT " YOU WOULD LIKE TO USE."
50      INPUT A,B
60      IF A > B THEN 80
70      LET C = A: LET A = B: LET B = C
80      FOR N = 1 TO B
90      IF N * A / B = INT (N * A / B) THEN PRINT
        " THE LEAST COMMON DENOMINATOR IS " ;N * A:
        LET N = B: GOTO 110
100     NEXT N
105     HOME
110     PRINT: PRINT: PRINT " AGAIN (Y/N)?"
120     INPUT AG$
130     IF AG$ = "Y"  THEN 40
140     END
```

```
THIS PROGRAM WILL HEI P YOU FIND THE
LEAST COMMON DENOMINATOR OF ANY TWO
DENOMINATORS.

TYPE IN TWO DENOMINATORS YOU WOULD
LIKE TO USE.
?2,5
THE LEAST COMMON DENOMINATOR IS 10

AGAIN (Y/N)?
?Y
TYPE IN TWO DENOMINATORS YOU WOULD
LIKE TO USE.
?12,8
THE LEAST COMMON DENOMINATOR IS 24
```

covery method. Students tend to refer to manuals not for sustained, lesson-by-lesson development but as references for solutions to specific programming obstacles that often require advanced techniques. By using this approach to instruct teachers, we can rightfully encourage them to teach others the way they were taught.

In summary, this discovery-based teaching strategy works very well with elementary school teachers and media specialists. Some reasons for this success have been suggested. I have been thoroughly impressed with the enthusiasm displayed by nearly all those taught by this approach. The general level of projects submitted has been very high. Teachers have wanted to create good, although not always short, programs and have worked very hard to do so. Perhaps nothing is more rewarding to a teacher educator than to see students enthusiastically embrace new, potentially intimidating ideas and then use these ideas creatively in educationally sound ways. The use of this strategy has given me this reward.

Figure 10 Problem solving: Combinations

```
10 REM COMBINATIONS OF *,<,/
20 DIM N$(3)
30 LET N$(1) = " *"
35 LET N$(2) = " <"
40 LET N$(3) = " / "
50 FOR L1 = 1 TO 3
60 FOR L2 = 1 TO 3
70 FOR L3 = 1 TO 3
80 PRINT N$(L1), N$(L2), N$(L3)
90 NEXT L3
100 NEXT L2
110 NEXT L1
120 END
```

```
*            *            *
*            *            <
*            *            /
*            <            *
*            <            <
*            <            /
*            /            *
*            /            <
*            /            /
<            *            *
<            *            <
<            *            /
<            <            *
<            <            <
<            <            /
<            /            *
<            /            <
<            /            /
/            *            *
/            *            <
/            *            /
/            <            *
/            <            <
```

They demonstrate the enthusiasm and curiosity characteristic of successful professionals.

Fourth, this approach to teaching programming is instructionally sound. It moves from the known (familiar information presented in print) to the unknown as teachers see the results (output) of executing BASIC constructs before they study the abstract meanings of the commands. This approach also makes possible a natural demonstration of the instructionally valid applications of computers in the classroom. Furthermore, teachers can take the listings of the programs with them as examples that they know work well.

Finally, this approach gives teachers a model to use when teaching programming to their own students. I have found that students, as well as teachers, learn elementary BASIC constructs well through this dis-

Teach BASIC Through Programming, Not Lecture

by Kenneth Kwajewski

The computer lab should be a place where a student can exercise freedom of the mind and not have to be subjected to a continuous lecture that may stifle the enthusiasm a computer can generate. Since a programming course may use a well documented text, though, it is easy for programming teachers to spend too much time lecturing on the idiosyncrasy of each command and statement contained within the pages of the book.

Instead, get the kids on the computers as soon as you can. You may be surprised at how fast they can comprehend the material without the need for lecture. Here's my approach.

Problem Solving

I actually call the course Problem Solving, but the guidance department keeps telling me I have to call it Programming for the college transcripts. Most of my students may never program a computer once

they complete my course, so I don't spend a great deal of time lecturing on every capability of every command. The first week of class is devoted to problem solving activities—outside, on the school ballpark. We play games that build class spirit and a desire to work together to solve a problem. One game we play is blindfolded soccer. The class is split into two teams, each team having half of its members blindfolded. The object is to have the students who aren't blindfolded give instructions to the sightless students so that they can kick the ball through the proper goal. This is a far cry from lecturing on every aspect of the print statement, and I think the time is well spent. We scatter many other activities that prepare students to work and think together throughout the school year in order to enjoy each other and take an occasional mental health day from constant programming.

Once the ice has been broken and the students realize this class is going to be enjoyable, I present material on programming commands and problem-solving methods. Most of this time is spent on an eight-step problem-solving method I borrowed from Vladimir Zwass' book, *Introduction to Computer Science.*

Eight Steps

The eight-step method has the following components:

- Problem definition;
- Problem analysis;
- Definition of data structure;
- Development of the algorithm;
- Program coding;
- Program implementation;
- Program documentation; and
- Program modification.

After providing the few necessary lectures on the programming language to cover most commands and statements, I show students how to solve a problem using the eight-step method.

The first step is problem definition. Students use the word processor to define the actual problem. The next step is to analyze the problem by stating what programming commands might be necessary to code the problem. Again, the word processor is used. Then the students look at different data structures that might be used to help handle the input and output.

In the beginning of the course this is easy to do, but as the year goes on more sophisticated data structures can be introduced. Algorithms can be described in terms of flowcharts or by outlining the solution in some type of pseudo code. The main focus is on first thinking the problem through rather than immediately generating code. Once the algorithm is approved, the code is written, tested, debugged, documented, and possibly later modified for a different application. You may have to constantly remind students to use this method, but it will work. It helps students learn how to solve a problem rather than worry about obtaining an answer. It also allows them to see how the computer can be used for other tasks besides programming.

Real Problems

From the very beginning I attempt to give my students problems that are real and therefore motivating. I do not use a textbook, but rely on handouts and lectures to provide material for the students. Once the initial lectures are completed they receive a packet of problems to solve from mathematics, science, social studies or any other area that is real to the student. Amortization tables, graphs and simple game programs are some of the early assignments. As students become proficient at programming, more sophisticated topics are presented. Lectures now are usually one day in length and happen only once every two to three weeks. There is, however, much discussion with individual students as they reach roadblocks in the eight-step process.

A major goal of the course is to complete an educational software project by the end of the school year. All assignments given earlier in the year help prepare students to devise a project that will be both useful and motivating. The entire fourth term is devoted to working individually on an assignment that will be used by other teachers or students in the school building. Examples of these projects are described later.

Individualize Instruction

Students will proceed at their own rates. Use this to your advantage by teaching commands and statements on an individualized basis. Advanced programming commands, sound and graphics can be taught very quickly in a small-group setting, and students are quite capable of picking up a manual

and learning material on their own. Quicker students will later teach other students. It is not necessary to stop every day and lecture on every topic; teach a new and then let the fire spread.

The Project

Let me illustrate how I used the above methods in one of our class assignments.

The year is only half over and we have covered variables, expressions, a host of programming commands, statements, functions, arrays, matrices, sequential and random files, graphics and sound. Did I lecture on all of these topics? No! They learned it the old-fashioned way; by doing it themselves or having it taught to them by other students or myself.

Our school is proud of its reputation for sending students on to higher education, and therefore many students are highly motivated to do well on the SATs. I suggested that the class develop an SAT program that would be used by students at our school to "brush up" for this examination. Here is how the eight-step method was employed.

1. Problem Definition

For a couple of days we talked about the problem and asked questions. What is an SAT program? Does anybody have one? Will the kids really use it once we create it? Can this program compete with commercial software? How should it be set up? Will it give a score? Should it be timed? Can the students save their scores? Should it have a review section? Who is going to do what? The answers to these questions helped guide us for the sections that followed.

2. Problem Analysis

Once most of the ideas were aired and discussed, it was time to talk about how the problem could be treated. More questions needed to be answered. Should it be a random examination? How should the menus be created? Should the student be allowed to interact with the review section? Should we use graphics or sound? How can we make graphs on the screen? The answers to these questions helped us to begin thinking about the programming job that lay ahead.

3. Data Structure

More questions had to be answered. Do we set up a file system? Should the whole program be print statements? How can gosubs be used with graphs? Do we use arrays to generate random questions?

Again the answers to these questions prepared us to write the program.

4. Algorithm

I had taught students how to flowchart and strongly insisted that they use it in their individual assignments. Some students who didn't like to flowchart were allowed to substitute an alternate type of algorithm. Due to the size of this project it was also necessary to create a class flowchart (see figure 1). Students could quickly see the need for the development of a plan of success.

5. Program Coding

As can be seen from the flowchart, each student had a specific programming assignment to complete. At times there were teams of programmers working on a specific module. Students could see the need for structure and coordination throughout the program. Natural cooperation between students evolved as individual talents were shared. I loved it because I became involved as well, and the students felt that we truly accomplished the project together.

6. Program Implementation

Students were excited to see their own finished programs, knowing that they would become part of a worthwhile project. I had every student examine another student's work to check for mistakes or to offer suggestions for improvement. As each module was completed, it was renumbered and added to the main program. The final program made every student proud of the work they had accomplished.

7. Program Documentation

The assignment had each student ensure that his/her program could be understood by other members in the class. It was obvious to every student that if the class had not used the eight-step method, the final program could not have been created.

8. Program Modification

How many times have you told your students that their program might be used again when in fact you knew that the chance was highly unlikely? In this example it was obvious to students that different modules could be added or subtracted at a later time. As students took the SAT review, it would become necessary to make different versions of the examination.

Could this project have been accomplished in most first-year programming classes? I doubt that it could have if the teacher tried to lecture on every statement that went into the program. But by using my method, students learned what they had to learn by using their own minds, manuals, other students or me to solve their individual problems. Isn't this how it is done in the real world?

This project was completed within the third term; the fourth term was then devoted to individual development of educational programs that were used throughout the building. Some of these projects included an inventory program for the library, a music tutorial program, a graphics tutorial program, a music generator program, a spelling program for middle school students. an educational mathematics game, an inventory program for the cafeteria, a file program to store statistics about the track team and a history exam program.

The SAT problem was not chosen from a book, but selected because it served a need. The students' motivation to solve this problem exceeded any enthusiasm that would have been generated by a textbook assignment.

Let Go

A programming course offers teachers the opportunity to not be obsessed with the bounds of a textbook; the computer does not have a cover, pages or page numbers. Don't confine students to their desks, but allow them the opportunity to use their creativity through programming. Let go of the reigns a little bit and let your students explore in a structured setting. Give students responsibility for their own learning. Let students teach other students, and you will usually find them extremely attentive to each other's ideas. Don't lecture them into passive submission, but let them exhaust themselves through creative and active programming.

The computer can become an object of discussion, or it can become a powerful tool for individual exploration. Which do you prefer?

Figure 1 Flowchart for SAT Program

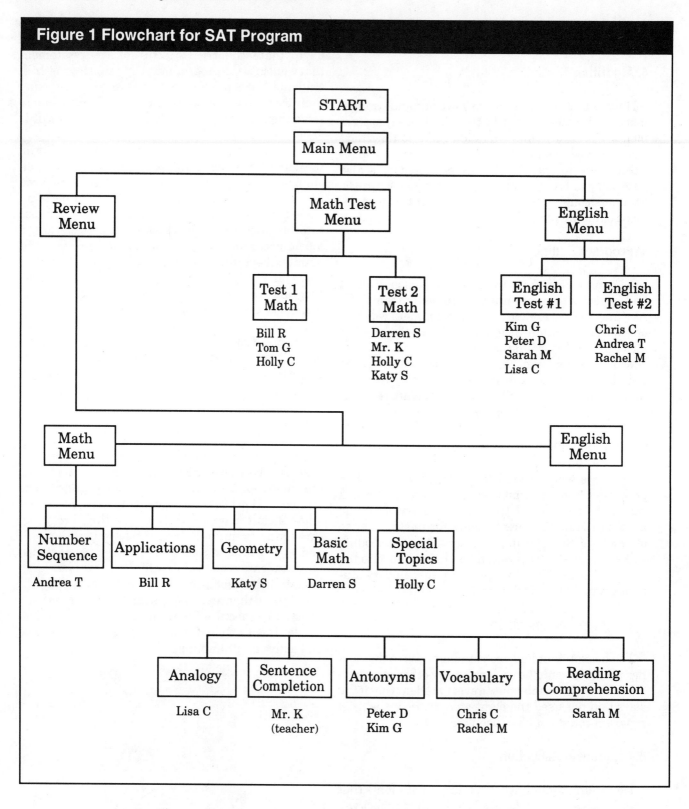

CHAPTER 7

Other Technologies in the Math Classroom

Computers are by no means the only technology of importance to the elementary mathematics teacher. Older technologies such as copiers, overhead projectors, film-strip projectors, and movie projectors have been and will continue to be used by teachers for a variety of purposes. However, older technologies such as film-strip and movie projectors are being replaced by the newer and more versatile technologies of videotape and videodisc players. The latter is a truly powerful device that brings a new level of imagery into the classroom, especially when it is used together with a computer to produce interactive video.

This chapter presents overviews of several technologies that are of importance to the mathematics teacher. First, and most important, is the electronic calculator. This device is changing the way we do arithmetic on the job and at home, and as a result it is also changing our mathematics curriculum. Of importance to the readers of this book is its use in problem solving. Telecommunications, though not as frequently used or as immediately important as the calculator, provides a powerful tool for communicating and accessing data. Interactive video will become increasingly important as good videodiscs for the mathematics classroom become available. Videotape is already an important tool. Many good mathematics videotapes/films have been produced. In addition, the teacher may use videocassette recorders to bring to the classroom television programs or materials he/she records from commercial television for the purpose of enhancing mathematics lessons. Finally, overhead projectors, although not an electronic technology, may be a powerful tool for enhancing mathematics instruction. Used with an overhead projection monitor, they can make the computer screen visible to an entire classroom of students.

Calculators and the Elementary Mathematics Curriculum

With the exception of the "new math" movement in the 1960s, the primary emphasis of elementary mathematics instruction during this century has been in developing (pencil and paper) computational facility. In grades 1 through 5, we spend huge amounts of time teaching children to add, subtract, multiply, and divide whole numbers. We start with addition and subtraction of one- and two-digit numbers in the first grade and by the fifth grade we have progressed to multiplication and division of three- and four-digit numbers. In grades 5 through 8, we shift our focus to computation with fractions, decimals, and percentages. Along the way, we teach other important mathematical concepts (e.g., the meaning of numeration systems and mathematical principles such as the commutative property). Other mathematical topics—problem-solving strategies, geometry, mathematical logic, the understanding of sets—have taken a back seat to the development of computational ability.

The existence and importance of calculators in our society have been recognized by the major organizations involved in mathematics education, such as the National Council of Teachers of Mathematics (1989), and the National Council of Supervisors of Mathematics (*Position Paper on Basic Skills*). Organizations involved in the designing of mathematics curricula such as the State of California (*see the Mathematics Framework for California Public Schools, Kindergarten through Grade Twelve*), are requiring curricular changes based on the impact of calculators. And elementary mathematics textbook series are changing to meet this challenge.

Some of the changes that are currently being made in the elementary mathematics curriculum as a result of calculators in our society are as follows:

1. Pencil-and-paper computations are de-emphasized and mathematical concepts such as numeration and the associative property are re-emphasized.

2. Estimation and mental arithmetic are specifically taught in school (rather than learned—or not learned—by individuals on their own). These skills are important for doing simple computations and for determining the reasonableness of computations done using technology.

3. Children are taught to use calculators and required to use them where appropriate.

4. Operations on fractions are de-emphasized while operations on decimals are emphasized. Of course, we still need to teach the meaning of fractions. But, since neither calculators nor computers can easily handle fractions[1], the emphasis is on decimals. Indeed, about the only place where fractions are regularly used is with nonmetric measurement. As we move more to the use of the metric system, the use of fractions will become even rarer. Of course, problems involving fractions still occur in daily life. However, any operation on fractions (e.g., 1/4 + 1/8) can be solved by converting the fractions to decimals, doing the operation with a calculator,

---

[1] There have been a number of calculators (e.g., the Texas Instruments TI-12) designed for the elementary mathematics classroom that do operations on fractions. About the only place a person will ever see such a calculator, however, is in an elementary classroom.

then converting the decimal answer back to a fraction. Children should know the decimal equivalents of commonly used fractions and they should be able to convert "odd" decimals to nearly equivalent, commonly-used fractions (e.g., 0.34268 is approximately equivalent to 1/3).

5. Children should be taught the meanings of decimal numerals at an early age (e.g., grade 2), as they will encounter them when they use calculators and in everyday life. Operations on decimals can be taught after students have mastered *simple* (two- and three-digit) paper-and-penciland mental algorithms as well as estimation techniques involving whole numbers.

Teaching Mathematics with Calculators

The calculator is, potentially, as important a tool as the pencil in the mathematics classroom. It can and should be used in developing mathematical concepts, in solving problems, and in checking answers obtained via other means. Of course, there are many times when it is not appropriate, for example, when memorizing "math facts" or when developing mathematical concepts using manipulative materials.

How should calculators be used in the classroom? Here are three suggestions:

1. *Doing computations in problem solving* and in any other situation where the objective is not to develop computational skill. For example, they should be allowed when adding a column of numbers obtained in a data-gathering activity, but not when children are practicing estimating the sums of columns of numbers, except to check the accuracy of the estimate.

In order to help you evaluate the processes the students are using when solving problems with a calculator (remember, you will not have their pencil-and-paper "scratch-work"), it is often a good idea to have them write the calculator keystrokes they use and intermediate answers they obtain. Fig. 7.1 shows a calculator problem-solving template that might be useful in encouraging students to write the calculator keystrokes as they attempt to solve problems.

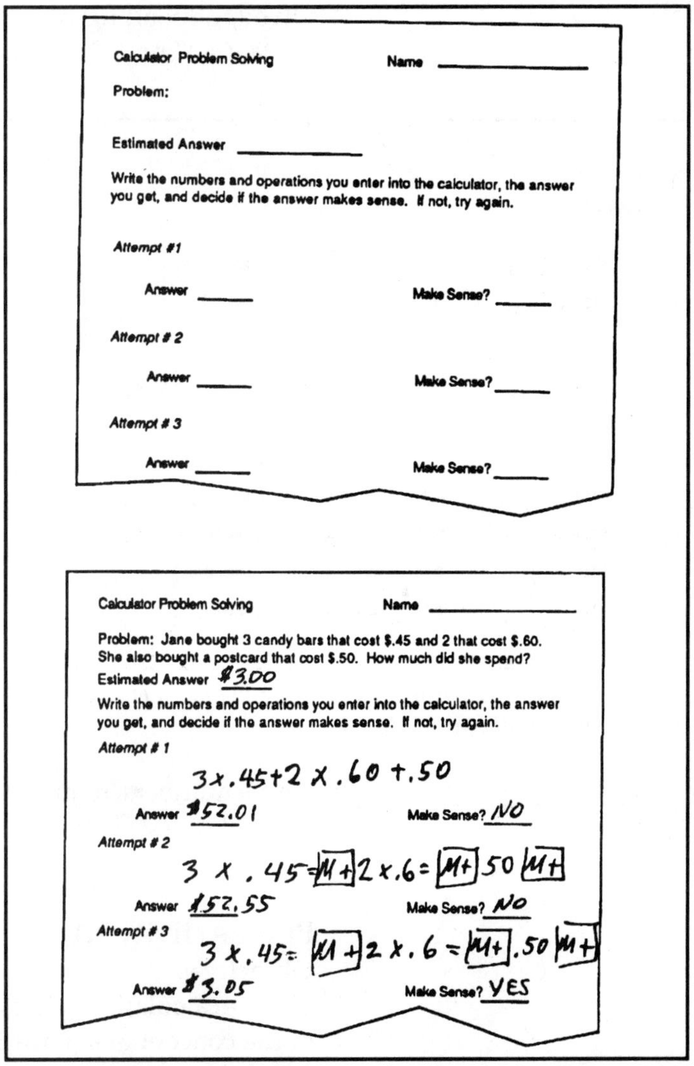
Calculator Problem Solving Name ____________

Problem:

Estimated Answer ____________

Write the numbers and operations you enter into the calculator, the answer you get, and decide if the answer makes sense. If not, try again.

Attempt #1

Answer ______ Make Sense? ______

Attempt # 2

Answer ______ Make Sense? ______

Attempt # 3

Answer ______ Make Sense? ______

Calculator Problem Solving Name ____________

Problem: Jane bought 3 candy bars that cost $.45 and 2 that cost $.60. She also bought a postcard that cost $.50. How much did she spend?

Estimated Answer $3.00

Write the numbers and operations you enter into the calculator, the answer you get, and decide if the answer makes sense. If not, try again.

Attempt # 1

3x.45+2 x .60 +.50

Answer $52.01 Make Sense? NO

Attempt # 2

3 x . 45 M+ 2 x.6 = M+ 50 M+

Answer $52.55 Make Sense? NO

Attempt # 3

3 x . 45 = M+ 2 x . 6 = M+ .50 M+

Answer $3.05 Make Sense? YES

Fig. 7.1: Calculator problem solving template (from Wiebe, 1988, pp. 38,39)

2. *Developing mathematical understanding*. The calculator is a wonderful tool for exploring whole numbers and decimals. Some examples follow:

A. The Constant Function: This can be used to help young children associate written numerals with verbal counting sequences, including backward and forward "skip counting" (skip counting prepares children for multiplication and division). For example, most calculators can be "pro-

grammed" to repeatedly execute the "+2" operation—adding 2 to the amount displayed—whenever the "=" button is pressed. With some calculators the "+2" operation is entered as follows:

$$+ \quad 2 \quad =$$

For other calculators, the following keystrokes are used:

$$2 \quad + \quad +$$

Now, each time the "=" is pressed, the "+2" operation is carried out. Once the "+2" operation has been entered, children can verbally count by 2s and observe the corresponding written numerals on the calculator's display. Another example is to have children solve multiplication problems by using the constant function to do repeated addition (e.g., 26 x 120 can be solved by entering "+ 120," then pressing the "=" button 26 times). Likewise, division problems can be solved using repeated subtraction (e.g., 2408 + 43 can be solved, depending on the type of calculator you have, by entering either "2408 / 43" or "43 - - 2408 " and pressing the "=" button until the display reads zero—the quotient is the number of times the "=" button is pressed).

EXAMPLE 7: √10 = ?

	Estimate		Trial		
STEP 1:	5	5	x	=	(25)
STEP 2:	4	4	x	=	(16)
STEP 3:	3	3	x	=	(9)

(√10 is between 3 and 4 because 10 is between 3^2 and 4^2)

Since 10 is closer to 9, one might estimate that √10 is close to 3.

	Estimate		Trial		
STEP 4:	3.1	3.1	x	=	(9.61)
STEP 5:	3.2	3.2	x	=	(10.24)
STEP 6:	3.15	3.15	x	=	(9.9225)
STEP 7:	3.16	3.16	x	=	(9.9856)
STEP 8:	3.17	3.17	x	=	(10.0489)

√10 is between 3.1 and 3.2

√10 is between 3.16 and 3.17

Fig. 7.2: *Using estimating and checking to find square roots (from Wiebe, 1988, pp. 35-36)*

B. Repeated Estimation: Since the calculator allows the user to do operations very quickly, it can be used for finding accurate answers through repeated estimation. After each estimate the student checks it with the calculator, then uses that information to make a more accurate estimate. Eventually, through estimation and checking, the user finds a precise answers. For example, this technique can be used to find square roots of numbers to several decimal places. To do so, try to find two consecutive numbers whose squares are larger and smaller than the desired number, progressing from whole numbers, to numbers with tenths, to numbers with hundredths, and so on. Fig. 7.2 shows one way to arrive at the square root of 10, accurate to two decimal places.

Pupil Activity 7.1
Objectives
Use the calculator to check computational estimates; develop understanding of the concept of square root.
Grouping
Individuals.

Materials
Calculator, pencil and paper.
Procedure
Go through examples of finding square roots to several decimal places as in Figure 7.2. Have students find several square roots using the same process.

Another example is to use estimation followed by multiplication on the calculator to develop meaning for division along with estimation ability. Students estimate the quotient of two numbers, then use the calculator to multiply the divisor by the estimated quotient to see how close to the answer they are. Students then make further estimates and check them until they find the exact quotient. Note that, when doing such activities, students should be required to write their estimates so that (1) you can verify that they actually estimated rather than simply dividing to find the exact answer on the calculator, and (2) you can observe the strategies and guide them to better estimation strategies.

Pupil Activity 7.2
Objectives
Estimate quotients, use calculator to check estimates.
Grouping
Individual.
Materials
Calculator, worksheet (see Fig. 7.3).

Procedures (Directions to pupil):
Estimate the quotients in the division problems in Fig. 7.3. Write your estimate, then use your calculator to multiply the estimate by the divisor. Write the product. Continue until your estimate is correct. Try to obtain the correct quotient within six guesses.

Activities like this not only help develop understanding of the concepts involved—square roots and division in the above two examples—but develop students' estimation abilities.

3. *Checking answers obtained by other means.* For example, if you had children do a set of estimations of sums of columns of multidigit numbers, they could use the calculator to find the exact sums, then assess the quality of their estimate (e.g., they could rate their estimate as "excellent," "good," "fair," or "poor").

Problem			Estimate	Answer
A. 5008/16	(1)	16 x	_____	_____
	(2)	16 x	_____	_____
	(3)	16 x	_____	_____
	(4)	16 x	_____	_____
	(5)	16 x	_____	_____
	(6)	16 x	_____	_____
	(7)	16 x	_____	_____
	(8)	16 x	_____	_____
B. 2310/35	(1)	35 x	_____	_____
		Etc.		

Fig. 7.3: Division by estimating and checking using a calculator

This chapter contains three additional articles about the use of calculators in the mathematics classroom. The article by Joan Spiker and Ray Kurtz discusses a variety of uses of calculators in the primary mathematics classroom. Katherine Willson, in her article, also discusses the use of calculators with primary-grade children, but focuses on problem solving. The article by Margaret Comstock and Franklin Demana gives suggestions for using calculators as a problem-solving tool with seventh and eighth graders.

One of the most positive benefits of integrating calculators into your mathematics instruction is that your program will emphasize *mathematics* rather than *computation*. At present, students tend to be overly concerned with obtaining a single, "correct" answer to every problem—and, in the shortest time possible. The calculator frees the student from tedious pencil and paper computation, allowing for experimentation, trying a variety of approaches (and evaluating each), and evaluating the reasonableness of answers.

Telecommunications

Telecommunications opens a classroom window to the outside world. It allows you, quickly and inexpensively, to send and receive messages to and from people around the world and to tap into information resources far beyond those available in your school. In mathematics, it allows you to expand the types of problems you bring into the classroom and also the resources available for solving problems. It provides a variety of resources and communications opportunities for you and your students.

Telecommunications resources available to the teacher

On-line services are commercial ventures that provide a variety of services to subscribers. Services such as *Compuserve* and *Prodigy* offer on-line shopping, weather from around the world, AP and UPI news, stock market information, hotel and rental car reservations, information data bases (e.g., article abstracts from computer magazines), and many, many other services. Compuserve and AppleLink both offer "forums" for teachers, allowing you to exchange ideas and information with other teachers around the country and the world.

One of the most important uses of on-line services is bulletin board systems for sending messages to and receiving messages from other subscribers. Depending on which service you subscribe to and your location, electronic mail services allow you to communicate with people around the world for the price of a local telephone call. Note that large services with large numbers of subscribers can be directly accessed in virtually all metropolitan areas in the United States. For example, *Compuserve*, with more than 500,000 subscribers, can be accessed via a local telephone call in more than 500 cities in the United States, meaning that about 85 percent of the U.S. population can log on directly, without paying for long-distance telephone rates. Compuserve is directly available in hundreds more cities and towns via carrier networks such as Tymnet, Telenet, DataPac, and others (Bowen & Peyton, 1989).

When you wish to send a message to someone via an on-line service, you enter a special editor where you compose your message. When the message is completed (and edited, if you desire), you enter the code or name of the subscriber to whom you are sending the message. The message is saved on the computing system of the on-line service and when the destinee signs onto the system, he/she is alerted that there is a message waiting. The person receiving the message can read it immediately or leave the message on the system for later reading. The message can be deleted as desired from the on-line system, saved to disk or printed on the receiver's own computing system, or even forwarded to another subscriber.

Most electronic mail services allow you to send and receive files (e.g., those created with a word processor) or computer programs. Thus, if you wish to send a long document, you can create it off line using your word processor, then send it after it has been completed (and edited!). Not only will this save you a lot of money (creating a 10-page document on-line might take two hours and cost you $15—most on-line services charge you according to the amount of connect time—while it might take two minutes and cost 50 cents to send the same document, created previously with your word processor), but you will be able to create and edit it at your leisure and using the powerful and familiar editing system of your own word processor rather than the unfamiliar and cumbersome editor of the on-line system. Note that during the past five years or so, most book and journal editors this author has dealt with required him to send manuscripts on diskette, rather than as printed copies. During the past year, however, he has experienced a new phenomenon: publishers are beginning to request that manuscripts be sent via telecommunications. Not only is this quicker, but the publisher does not need to try to find a computing system and software that matches the author's—it can be received and saved by their own system, whatever make it is.

There are a number of other on-line services of specific interest to teachers. *Knowledge Index* provides after-hours access to *Dialog*, the computerized data base used by college libraries. This service allows you to research topics of interest to you or your students. For example, through Dialog, you could access the ERIC database and get lists of journal article references and abstracts about using calculators in teaching problem solving in the elementary grades, or articles describing activities using spreadsheets in the mathematics classroom. *Knowledge Index* provides additional services to educators such as *Books in Print*, a Microcomputer Software Guide, Peterson's College Database, the Pollution Database, UPI News, Facts on File, and many, many more.

There is a variety of bulletin boards offering services to teachers and students. FrEdMail is a computer network with dozens of outlets around the United States and the world. It provides information for teachers (e.g., lesson plans, teacher services) and allows children to communicate with each other. *KidsNet* is a children's network supported by *National Geographic*. It provides opportunities for children to work collaboratively from other parts of the world to solve scientific problems.

Student Use of Telecommunications

Children find communicating with students in other school districts, towns, and countries via the computer to be very exciting. There are several advantages of telecommunications over face-to-face contact. Most important, it makes quick, "direct" communication affordable. Telephone conversations with students in distant locations would be prohibitively expensive for most schools. In locations where FrEdMail is available, students can send messages to children in distant locations without having to dial long distance. The same is true if the teacher or school is willing to pay the price of membership in Compuserve, AppleLink, Prodigy, or another on-line service. And, of course, these commercial services are available everywhere, not only in select locations, as are "voluntary" networks and local bulletin board systems.

An important benefit of telecommunications is that, when communicating via keyboard and computer screen, the stereotypes children have based on appearance, accent, and the like disappear. Children do not categorize the person with whom they are communicating as "handicapped," "ugly," "male," or "female," but, rather, react to the content of the message received.

Another benefit is the excitement of using computers as tools for communication. Whereas many students find letter writing by hand to be a drag, they will compose long messages on the keyboard with great enthusiasm. And, finding a message waiting on the computer bulletin board is very exciting.

Here are two suggestions for the use of telecommunications in the mathematics classroom:

1. Using on-line data bases and information sources in solving problems. The services available on *Knowledge Index* or *Prodigy* can be used in solving problems like some of those suggested in Chapter 2 and 3 (e.g., Pupil Activity 3.5).

2. Create Interschool cooperation in solving problems. Telecommunications offer a number of possiblities in cooperative problem solving. One possibility is for teachers in two different schools to arrange to have their pupils write problems, then send them to the other class to solve. Each group would solve the problems, and send back the answers. Each group then would verify the correctness of the answer.

Another possibility is to create problems that require information that students obtain from children in another class, via telecommunications. An example would be to gather information of interest about the children in the other class (a similar, but shortened version of the questionnaire in Pupil Activity 3.2) via telecommunications, then evaluate it using a spreadsheet or data base. At the same time, the other class could develop their own questionnaire and obtain and evaluate the information in the same way.

Pupil Activity 7.3 Weather Data/Telecommunications

Objectives
Collect and evaluate weather data in metric system; use telecommunications for exchanging information; graph weather data.

Grouping
Whole group.
Materials
Celsius thermometer, rain-collecting device, metric ruler.

Procedure
Locate a cooperating teacher in a location with a different climate from yours (e.g., FrEdMail allows you to post messages that any subscriber can read).

Set up data-gathering activities involving the weather in the metric system. For example, place a Celsius thermometer in a shaded area outside your classroom and have the children record the temperature at 8:30 AM and 3:00 PM every school day during the month of February. Also, place a cylinder in an open area where it will not be molested by people—e.g., on the workshed roof in the fenced maintenance area—and measure the amount of precipitation in centimeters at 8:00 every morning. If the precipitation is in the form of snow, melt it first. Send the weather data every Friday to the other group. Enter the data gathered at your

Week	Feb 5, 1991	Feb 12, 199	Feb 19, 199	Feb 26, 199	Mar 5, 1991
SD High	71	69	62	69	67
SD Low	42	41	49	41	43
Chi High	24	54	48	32	31
Chi Low	15	28	22	18	5

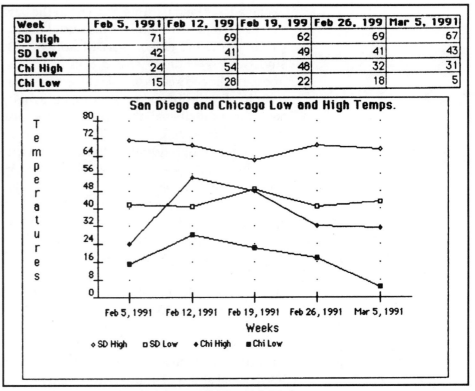

Fig. 7.4: *Sample weather data done with Microsoft Works*

school and at the other school into a spreadsheet. Graph the results (e.g,. you could graph the average low and high in each location for the four or five weeks of the experiment. Use the resulting graphs to help make observations about the differences in weather in the two locations.

Telecommunications could provide the opportunity for friendly competition between classes in different locations. Memorizing the "basic facts" (single-digit operations such as "5 + 6" and "7 x 8") continues to be an extremely important part of the curriculum. Competing classes could hold timed tests (exchanged by the teachers using telecommunications at the beginning of the week) every Friday over the same skills. Each class could compute the average score on the test and send the results to the the other class using electronic mail. At the end of the competition, the winning and losing classes would exchange congratulations or condolences. The classes could also exchange student-created math/computer art, cards created with Print Shop, Logo designs, and the like.

Interactive Video

Compact disc technology has taken the recorded music business by storm. Not only is the sound quality of digitally recorded music superior to that of music recorded on other media, but the built-in computer components in the more sophisticated players allow you to program your player to repeat a track a given number of times or play selected tracks on the disc. Furthermore, the discs themselves are virtually indestructable since the codes containing the music are recorded and read from the *interior* of the discs using a laser beam, rather than from the surface.

Laser-disc technology is also used to store and play video images. The larger (8 1/2 in.) videodiscs can store over 100,000 high-resolution images and two high-fidelity sound tracks. This is sufficient to store full-length movies. Videodiscs are available in two formats. The CLV format allows the user to play the images on the disc in sequence, as moving pictures. Images are stored on all available spaces on the disc tracks, thus allowing for 120 minutes of playing time. The disadvantage of this format is that the displaying of individual (still) frames on the disc is not possible.

The more powerful and flexible CAV format, however, allows the user to access and display single images, or play the frames in sequence as a movie. These images can be quickly accessed in any order—in fact, the maximum scan time for moving from one image to another on the disc is about 2 seconds. In the motion-picture format, however, there is a disadvantage to CAV: because of the way images must be stored to allow single-frame viewing, each disc can contain only half as much as in the CLV format—about 60 minutes of continuous playing time.

Discs in the CAV format designed for use in education often hold both individual "slides" and motion picture sequences. Note that individual frames of the motion picture sequences can also be accessed and played as still images. Thus, a videodisc about jungle animals could contain both individual slides of animals and motion pictures of their movement, hunting/eating habits, and mating rituals. The laser-disc players can be controlled either by a set of buttons on the player, by a remote controller, or via a computer.

There are many educational discs available, especially for use in the arts and sciences. For example, there is a series of astronomy discs with photographics of planets and star systems. There are various discs containing photographs and movies from various art galleries or which contain the works of particular artists. There are also numerous discs designed to supplement instruction in various subjects. In mathematics, the Mastering Fractions disc (Systems Impact Corp.), is designed to help children understand fractions. The number of educational discs being published has increased dramatically over the past few years.

Using laser-disc players in the stand-alone mode to show still or moving images to students is an exciting new way to bring the outside world into the classroom. An even more exciting use of this technology is to couple it with computers to produce interactive video. With appropriate interfaces (plugs and cables), computers can be used to control the video player. This means that the teacher can set up a program—designed with any number of simple tools such as Hypercard, Hyperstudio, Linkway, or AppleWorks[2]—to access the desired frames on the videodisc. This also means that he/she will not need to look up and enter four or five-digit codes into the remote controller to access the frame he/she wishes to show. Many new Hypercard stacks or instructional programs created with authoring languages now are designed around laser discs, bringing a much higher level of graphics and sound to computer-assisted instruction. Many of these tools are friendly enough that teachers can, with a minimum of study, create their own instructional sequences with prerecorded laser disks.

Until recently, the equipment needed for interactive video was quite expensive—upward of $2000 for a videodisc player, a monitor capable of displaying the output from the player (most computer monitors are incapable of displaying the output from laser-disc players), and the needed cables and interfaces. That cost, however, has been reduced significantly. Apple Computer Corporation now sells inexpensive video cards that allows you to display the output from the laserdisc player on an Apple IIgs or color Macintosh monitors. Thus, there is no need to purchase an additional monitor for displaying the output from the video player. Also, with a one-monitor system, children are not confused by the presence of two monitors, one for output from the computer and one for output from the videodisc player. In a one-monitor system, the text (and graphics) from the computer and pictures from the videodisc are intermixed as the software author desires.

You will be hearing much more about interactive video in the near future. Business and the military are spending large amounts of money to develop laser-disc/computer software to train their personnel. And, in the near future, you will be able to purchase computers that incorporate videodisc players in the same unit—intended for educational use—at reasonable costs. For the math teacher, there will be a variety of software designed to help students understand mathematics concepts and solve problems.

Videotape

Consumer-oriented videotape and videodisc technology appeared at about the same time. But, because people could record broadcast television pro-

---

[2] Macros and instructions for using AppleWorks to control videodisc players are available through Teaching Technologies, P.O. Box 3808, San Luis Obisbo, CA 93403

grams for later viewing and could use video cameras to make their own movies, videotape technology florished while videodiscs languished in the consumer market. Thus, players, cameras, and software (e.g., movies recorded on videotape) became relatively inexpensive. The advantages of laser videodisc—higher resolution video images, very quick access to still images, duel stereo sound tracks—were overlooked in favor of the recording possibilities of videotape.

Despite the lower quality of the images on videotape and the lack of quick random access to single (still) frames, videotape teachnology has a variety of uses in the mathematics classroom. One of the most important uses is to record instructional programs from broadcast television such as "3-2-1 Contact," to show in the classroom. Another possibility is to record segments of commercials or entertainment programs that contain mathematical content or from which data can be gathered for data-gathering activities.

Likewise, a video camera can be used to bring images into the classroom for applied mathematics activities or problem solving. For example, if you decided to do graphing activities with first- or second-graders involving the different colors of cars passing by a particular spot or the number of people with different hair colors, it would be quicker, easier, and safer to have a camcorder record the event than to take the entire class to that location. Another advantage of using a camcorder to record these events is that you can replay the tape as often as needed so that the appropriate information can be viewed and tabulated by the children.

Another use of camcorders is to represent problems to enhance children's understanding of them. Problems involving people, money, physical objects, and such could be represented or acted out, recorded, and played for (or by!) children to help them understand the problem situation and begin to devise solutions.

Overhead Projectors

Manipulating physical objects to represent mathematical concepts to children is an essential step in developing understanding. Trying to manipulate counting objects, numeration blocks, number lines, and the like for an entire class to see, however, is difficult, if not impossible. Drawing pictures on the chalkboard or asking children to look at pictures in their textbooks will be of little help, since these representations are static (e.g., they cannot show the actual *joining* of two sets of objects to model addition).

The overhead projector makes it possible for the teacher to demonstrate mathematical concepts using objects that all can see. You can use actual objects (e.g., dried garbanzo beans, poker chips) or pictures of manipulatives drawn on transparencies (e.g., numeration blocks, fraction shapes, number lines). See the article entitled "Teacher-made Overhead Manipulatives" at the end of this chapter for detailed instructions on making manipulative materials for the overhead projector.

Another problem solved by the overhead projector is the difficulty of seeing the contents of small computer screens in classroom demonstrations. Transparent LCD computer monitors can be attached to your computer, placed on the overhead projector, and projected on a screen so that the output is visible to a large group. Low-end overhead projector monitors are now available for less than $500. Most models are designed for specific types of display standards (e.g., composite color, IBM-CGA, Macintosh mono-chrome, etc.). Monitors that can be used with different output standards, high resolution, and color projection monitors are more expensive.

Another useful device is a calculator designed to be used on an overhead projector. They are inexpensive (e.g, *The Educator* is available from Dale Seymour Publications for less than $50) and allow the entire class to see the display. You should consider purchasing an overhead calculator, such as *The Explorer*, which not only has a transparent display, but also transparent keys so that your students may observe the keystrokes you make. Overhead calculators are useful for teaching children to use calculators. They may also be used in a variety of activities involving estimation. For example, the teacher gives the class a problem such as "405 x 21" and asks them to estimate the answer and write it on paper. The teacher does the computation using the overhead calculator. The class then evaluates the estimated answers and their method for obtaining them and discusses the accuracy of the estimated answers. Exploration of number patterns (e.g., patterns using the constant function) and realistic problems involving large numbers or numerous computations are also aided by the overhead calculator.

New technologies that may have an impact on the classroom appear regularly in our technologically oriented society. Some of these—such as television and computers—have been touted by some as having the poten-tial to revolutionize education. Although no revolution is immediately on the horizon, these innovations, especially the computer, are tools that are enhancing traditional instruction in a variety of ways. Teachers who use them will add excitement to their classroom, and at the same time arm their students with powerful tools for learning.

Bibliography

An Agenda for Action: Recommendations for School Mathematics of the 1980s. 1980. Reston, Va.: National Council of Teachers of Mathematics.

Arithmetic Teacher 1987. Special Calculator Issue 34(5, January).

Bitter, G. G. and Mikesell, J. L. 1980. *Activities Handbook for Teaching with the Hand-Held Calculator.* Boston: Allyn and Bacon.

Bowen, C. and Peyton, D. 1989. *How to Get the Most out of CompuServe: Fourth Edition.* New York: Bantam Books.

Curriculum and evaluation standards for school mathematics. 1989. Reston, Va.: National Council of Teachers of Mathematics.

Encyclopedia of Animals. Pioneer Laserdisc Corp. (1-800-255-2550).

First National Kidisc. Optical Programming Associates (available through Ztek Co. 1-800-247-1603).

Mathematics Framework for California Public Schools, Kindergarten through Grade Twelve. 1986. California State Department of Education.

Miller, D. 1979. *Calculator Explorations and Problems.* New Rochelle, NY.: Cuisenaire Co. of America.

Position paper on basic mathematical skills. (1977) National Council of Supervisors of Mathematics, *Arithmetic Teacher*, 25 (2, October) 18-22.

Verport, G. and Mason, D. 1980. *Calculator Math.* Fearson Teaching Aids.

Wiebe, J. H. 1988. *Teaching Elementary Mathematics in a Technological Age.* Scottsdale, Ariz.: Gorsuch, Scarisbrick.

Wiebe, J. H. 1990. Teaching mathematics with technology: Teacher- made overhead manipulatives. *Arithmetic Teacher* 37(7, March): 44-46.

Discussion Questions

1. How can videotape be used to enhance mathematics instruction?

2. What are some of the advantages of the use of laserdisc for video output as opposed to traditional computer graphics?

3. How has the calculator changed the way you do mathematics? How have calculators, computerized cash registers, and computerized checkout (using bar code readers) changed the types of math people must do in the business world?

Activities

Activity 7.1
Objective
Develop lesson plans that incorporate technology into mathematics instruction.
Rationale
Activities involving technology should not be isolated, supplementary activities: rather they should be fully integrated into the your curriculum. Developing appropriate lesson and unit plans that focus on the content to be taught will help in this process.
Materials
Pencil and paper, calculators, cassette player/recorder, or television set.
Advance Preparation
Ability to do basic operations on a calculator; knowledge of mathematics curriculum and the use of the calculator in teaching/practicing mathematical concepts; use of videotape player for recording and playback of commercial television broadcasts.

Procedure
1. Develop an activity similar to the estimation/division activity, discussed in this chapter, for estimation and subtraction of whole numbers, using calculators.
2. Develop a lesson plan involving the use of programs taped from commercial television in the mathematics classroom.

Activity 7.2

Objective
Use of a laser-disc player.

Rationale
Videodisc is a powerful tool for bringing the outside world into the classroom and for presenting instructional material to children.

Materials
Laser-disc player with handheld controller; educational laserdisc software suitable for use in math or integrated math activity; a computer interfaced with a laserdisc player; authoring software capable of controlling the laserdisc (e.g., Hypercard with a videodisc-controller stack, Hyperstudio, Linkway, etc.)

Advance Preparation
Learn how to control the laser-disc player using the handheld controller. Learn to use the computer authoring system and its functions that control the laserdisc player. (*Note:* You will not use the authoring system to create computer-assisted instruction, only to access the desired frames on the laser disc.)

Procedure
1. Select an educationally oriented laser-disc title (e.g., *Mastering Fractions, First National Kidisc, Encyclopedia of Animals*). Go through the directory of the disc and select a set of frames and video sequences that you would like to view. *Note:* These should be scattered throughout the disc. Use a handheld controller to access and view them. Discuss how the disc could be used to enhance instruction in your class.
2. Now, use the authoring software to reproduce the sequence of frames and video sequences: *(a)* Set up a program so that you can control the laser-disc player by pressing keys on the computer to step through the sequence of frames (e.g., each time you press the space bar, the player moves to the next frame on the list), and *(b)* create a menu with the authoring software to jump to any of the frames or video sequences on your list.

Teaching Primary Grade Mathematics Skills with Calculators

by Joan Spiker and Ray Kurtz

When colleagues see calculators in my first grade classroom they ask, "How will they be used in the first grade?" Professional pride dictates that I come up with the best answer possible. I tell them I am going to teach (1) what calculators are, (2) what calculators do, and (3) how calculators are used. Children should become comfortable using these tools. The goal is to teach and reinforce the objectives currently required in the first grade curriculum.

Since the calculator is one of the tools of our era, it should be introduced in the primary grades. Using calculators in the classroom offers children the opportunity to grow up with technology. The four-function solar-powered calculator is an ideal choice because it is inexpensive. It is important to place calculators in the hands of every student to involve them in technology. In this way they do not have to wait to participate, which is counter to most small-group models used in K-2.

The activities in this article have been developed and used in a first grade classroom with twenty-eight children, one teacher, and a teacher's aid. The calculator activities proved to be an exciting component of the standard curriculum. The children were thrilled to use calculators. Mathematics time became more interesting for the children when they knew calculator activities were to be done. These were a unique supplement to textbook mathematics.

In covering the first topic, "What calculators are," we discussed the needed vocabulary. The word calculator was written on the chalkboard, discussed, and repeated in unison. The cover of the calculator was opened and the light-powered solar calculator was described. The keyboard was identified. The children were shown how to enter their telephone numbers (without dashes) and how to clear the display.

A class discussion was held of the second topic—"What calculators do." Even though the children knew what calculators were, very few had a grasp of what they do. One said, "My dad has one in his checkbook." Another student said, "My mother used one when she shopped for groceries." The class discussed how the calculator would add, subtract, and "remember" lots of numbers. The children were encouraged to find out other possible uses and report back.

The third topic, "How calculators are used," took most of the classroom time. Activities were designed to teach and reinforce listening skills; the meaning and use of numbers, reading of number words, application of the "greater than," "less than," and place value concepts; and problem solving. Many of the activities involve worksheets that can be used with individuals, small groups, or large-group instructional settings.

Introductory Activities

Keying

Distribute calculators and introduce the following vocabulary words: *calculator*, *keyboard*, and *display*. Do location exercises with a large keyboard chart. Locate each number, key it in, and clear. Practice entering numbers as shown of the fingering chart (fig. 1). Point out that pencil erasers should not be used to key in because the erasures will fall into the keys. Use of the eraser can be compared to the use of only one finger in typing. Locate the function keys +, -, and =. Practice some basic addition and subtraction problems. Allow students to become familiar and comfortable with the calculators.

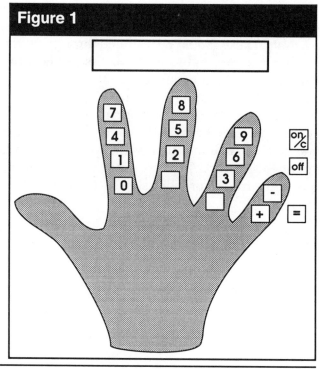

Figure 1

Enrichment: Dictate random numbers from 1 to 100 to be keyed in by the students. To check, point to a large hundred chart or write the number on the chalkboard. Try numbers beyond 100.

Electronic addition and subtraction

Practice calculator skills with teacher-made worksheets on simple and difficult addition and subtraction problems.

Enrichment: Student should write their own problems for addition and subtraction practice and use the calculator to solve them. Encourage students to use multiple addends in addition problems, for example:

3 + 3 + 2 = ?

or

7 + 1 + 2 + 0 + 4 + 3 = ?

Patterns

Demonstrate skip counting using the = key. Record the patterns on a grid:

1 + 1 = = = = = = = = = = (by ones)
2 + 2 = = = = = = = = = = (by twos)
5 + 5 = = = = = = = = = = (by fives)
10 + 10 = = = = = = = = (by tens)

Enrichment: Have the students work in pairs and experiment with other patters, such as 100+100. Record these patterns on a large poster for all to see. The title of the poster might be, "Wow! Look at These Patterns!" These posters generated a great deal of enthusiasm and set up an important link between home and school when parents and siblings became involved.

Skill Lessons

The order of things

Objective: To develop the concept of the communicative property of addition.

Sample activities:
Solve these problems:

3 + 1 = ?
1 + 3 = ?
9 + 2 = ?
2 + 9 = ?
7 + 3 = ?
3 + 7 = ?

Then construct similar equations:

? + ? = ?
? + ? = ?
? + ? = ?
? + ? = ?
? + ? = ?
? + ? = ?

Enrichment: Experiment with construction of more difficult equations:

500 + 100 = ?
100 + 500 = ?
200 + 100 = ?
100 + 200 = ?
600 + 300 = ?
300 + 600 = ?

Greater than or less than

Objective: To develop the concepts of greater than and less than.

Sample activities:
a. Use the calculator key sequence +1 to find a number that is one greater than each of the following:

72__   99__   81__
19__   35__   40__

b. use the codes +1 and - 1 to find a number that is one more than and one less than each of the following:

__29__   __94__
__75__   __31__   __100__

Enrichment: With the same format find numbers that are two more than and two less than a given number, five more than and five less than, and so on. Move on to more difficult numbers, such as these:

__324__   __710__

<u>995</u> 1000 <u>1005</u> (example)

A challenging variation is to give the three numbers and ask students how to get from one number to the next one: 705 <u>710</u> 715. (Answer: five less than and five greater than 710).
  Circle the larger number in each pair. Use the calculator to add the three circled numbers. Check your sum.

6	9
3	1
4	3

Circle the smaller number in each pair. Use the calculator to add the three circled numbers. Check your sum.

4	7
32	30
9	11
43	

Guess and check

Objective: To develop the guess-and-check strategy.

Sample activities:
Supply the correct sign, + or -:

7 _ 3 = 10   6 _ 6 = 0   4 _ 5 = 9

Supply the missing addends

3 + ? = 20   7 + ? = 15   ? + 4 = 10

Circle three numbers with a sum of 11:

7   2   1   3

Circle three numbers with a sum of 19:

8   4   3   8

Circle three numbers with a sum of 12:

3   4   7   2

Enrichment: Use the guess-and-check strategy to determine your answer. Circle it. Then check your answer on the calculator.

The sum of 29 and 10 is about

20   30   40

The sum of 50 and 18 is about

50   60   70

The sum of 41 and 30 is about

60   70   80

Number sense 1

Objective: To develop the foundation of estimation (counting on).

Sample activities:
Find more than two addends to make combinations for these numbers:

3 + 1 + 9 + 5 + 0
1 + 6 + 1 + 1
42
100

Enrichment: Try starting at the first number and count down four times with your calculator to reach the last number.

100 - 30 - 20 - 10 - 8 = 32
300 - ? - ? - ? - ? = 50

Problem solving

Objective: To develop problem solving. Calculators allow the students to focus on a process in story problems, which alleviates computational frustration.

Sample activity:

Construct story problems and solve them with the calculator:

Jim had an aquarium with 7 goldfish, 21 guppies, and 3 angelfish. The guppies had 20 babies. How many fish were in the aquarium?

Susan had $5.00. She then received an allowance of $4.00 and earned $1.00 helping her mother clean the house. Would she be able to buy a record that costs $7.95?

Enrichment: Play "store" using calculators as cash registers. Use actual items, catalog pictures or grocery ads

Ordinals

Objective: To develop the concept of ordinal numbers.

Sample activity

Add the first, third, and sixth numbers in these rows: Check your sums.

10   9   3   0   20   15   4   8   <u>28</u>

40   11   6   1   10   2   0   7   <u>48</u>

Number sense 2

Objective: To develop number sense in addition and subtraction.

Sample activity

Key in 35. Add two tens. Now you have __.

Key in 57. Take away all the tens. Now you have __.

Key in 19. Add four tens. Now you have __.

Key in 99. Take away all the ones. Now you have __.

Number words

Objective: To help students recognize number words.

Sample activity

With calculators and cards with number words on the front and numeral words on the back, students read the name, key it in, and turn the card over to check.

| nine | -> 9 | seventy-two | -> | 72 |
| forty | -> 40 | eighty-four | -> | 84 |

Summary

The question facing primary-school educators is not "Should we use calculators?" but "How should we use calculators more efficiently and more enthusiastically to meet the objectives of the existing curriculum?" Calculators should not be thought of as a shortcut or crutch or used in lieu of children's learning the basic facts. Rather, they should be used in developing computational skills. Most important is that by using calculators, children will be at ease with technology and be able to move into the computer age.

Calculators and Word Problems in the Primary Grades

by Katherine J. Willson

One teaching strategy for problem solving that appears to be highly successful is to have pupils write their own word problems. This process involves pupils in a teaching method that combines reading, critical thinking, and examining data (Fennell and Ammon 1985). Some pupils, however, may write problems containing numbers that require computation beyond their capabilities. This situation requires that students use calculators to solve problems.

I tried this approach with a first-grade class. The class was asked to create a problem-solving booklet to which each pupil contributed the favorite of the problems that she or he authored. We emphasized solving real problems found in the classroom, school, or home environment. Many teachers and researchers are aware that students are highly motivated when asked to solve problems that are "real" to them as opposed to contrived problems found in the textbooks. Trading stickers and hockey cards was the current rage, and many pupils were enthusiastic about writing problems that dealt with trading.

The pupils understood the meaning of addition and subtraction and throughout the year had solved word problems involving basic facts. They were able to read and write numbers to 100 and had learned basic facts to 10. As this experience was new for the pupils, we began with a discussion in which they created oral story problems that were recorded on the chalkboard. The intent was not to solve each problem as it was recorded but to brainstorm many oral problems so as to explore the kinds of problems that the pupils would create.

Solving Class Problems

To determine whether the pupils could solve problems from the brainstorming session, five problems from a collection of fifteen were presented to the class. The pupils were told that some problems might be difficult to solve but not to worry if they were unable to solve a problem because the class would have the opportunity to solve them together at the end of the period. As each problem was written on the chalkboard and read aloud, the pupils were asked to write a number sentence for the problem and then to try to complete the number sentence by solving the problem any way they could. Each pupil had a container of approximate 100 beans in his or her desk, and Unifix cubes were available at the back of the room. The pupils had used both manipulatives extensively while learning many new concepts, including calculations with two-digit numbers. They worked individually and finished one problem before moving to the next.

The class seemed enthusiastic. After each problem was read aloud, I circulated to observe the pupils and noted the strategies they chose to solve each problem. Although many pupils correctly identified the problems as addition or subtraction situations and wrote the initial number sentence correctly, few of them correctly completed or solved more than two of the five problems. Following the individual problem solving, the class shared their solutions and strategies. Examples of the pupils' attempts to solve four problems and the strategies that they used for each problem are as follows:

Problem 1

Today we played hockey at recess. There were 10 players on Michael's team and 7 players on mine. How many more players did Michael's team have than my team?

The majority of pupils recorded and completed a correct number sentence for the first problem. Most of them appeared to solve the problem through mental calculation, and a few used their fingers as counters. When queried, most pupils said that they had solved the first problem in their head. Some stated that they knew that seven plus three was ten, and others stated that they counted backward from ten to find the answer. When asked if the problem could be solved in any other ways, pupils suggested "drawing pictures," "using fingers," and "using counters."

Problem 2

One day I had 44 stickers. My dad gave me 18. How many do I have now?

This problem caused some concern. Pupils had little difficulty recording the correct number sentence, $44 + 18 = ?$; many, however, appeared confused when trying to solve it. Several used counters, and a few started drawing pictures to represent stickers. Those who tried to solve the problem mentally were unsuccessful. Most of the class attempted to solve the problem and completed the number sentence. However, only about a quarter of these pupils arrived at the correct answer. Those who solved the second problem said that they counted 44 beans and then 18 beans and added the two piles of beans together. Some pupils counted the combined pile starting with 1; others counted on from 44. A few pupils used their knowledge of place value to group counters in tens and ones and then added the groups together. Alternative strategies that the class suggested would be feasible to solve this problem were "act out the problem using real stickers" and "use a calculator."

Problem 3

Carl had 50 cents and he found 25 cents in the hall. How much money does he have?

The majority of pupils recorded 50 + 25 = ?, but many did not attempt to complete the number sentence or solve this problem. A few pupils struggling with counters counted a group of 50 and then a group of 25. Approximately a quarter of the class, of which the majority appeared to compute the answer mentally, correctly solved the problem. Those who solved the problem correctly stated that they were familiar with money. They said they knew that "two quarters, or fifty cents, plus another quarter made three quarters, or seventy-five cents." The only other strategy the class suggested to solve this problem was "act it out using real money."

Problem 4

One day a girl had a computer. She had to do 100 mathematics questions before doing the other 200. When she was done with the 200, how many did she do altogether?

Many pupils appeared discouraged or frustrated with their inability to compute using large numbers. Only three attempted to record a number sentence for the last problem. All three appeared to compute mentally to find the solution, and two of them correctly solved the problem. The two successful pupils said that they knew that 100 and 200 were 300. The only alternative strategy the class suggested to solve this problem was to "use a calculator." The pupils explained that counters would be too difficult to use because they did not have enough in their desks. Others responded that they did not attempt to solve the problem because "the numbers were too high," "too big," or "too hard to add."

Introducing the Calculator

The class problem-solving session led me to realize that although the pupils appeared eager to create problems involving large numbers, they quickly became discouraged when attempting to solve these problems. Even those who correctly used counters found them time-consuming and cumbersome, so they quickly became discouraged. How was the dilemma solved? During our class discussion, the pupils had suggested using the calculator as a viable strategy for solving problems. This alternative was not particularly surprising, as all pupils were familiar with the calculator and had used it throughout the year. They had been introduced to calculator activities from Keystrokes (Reys et al. 1980), which explored counting and place-value concepts. Students had used the calculator to count on, count backward, skip count, look for patterns, and add and subtract basic facts. Therefore, at this point I decided to teach students the mechanics of adding and subtracting two-digit numbers on the calculator to see what effect calculator use would have on problem solving.

Initially, such manipulatives as Unifix cubes were used in conjunction with the calculator. To promote an understanding of adding and subtracting two-digit numbers, we began computing with multiples of ten. For example, a number sentence, such as 20 + 10 = ?, was written on the chalkboard. To stress estimation, the pupils were asked to predict the answer and record their prediction. They then solved the number sentence by using Unifix cubes grouped in tens. They confirmed their answer by using the calculator. When pupils began working with numbers that were not multiples of ten, rounding to the nearest ten was taught. Although some pupils had difficulty rounding to the nearest ten, many were able to determine if the calculator answer was reasonable. For example, when given the number sentence 46 + 37 = 43, most pupils agreed that the number after the equals sign could not be 43 and was in fact close to 70 or 80. Many pupils estimated 70 by using front-end estimation, truncating 46 to 40 and 37 to 30. After adding 46 + 37 with a calculator, most agreed that 83 was close enough to 70 to be reasonable or to make sense. Before long, the majority of pupils became proficient in the mechanics of adding and subtracting two-digit numbers using a calculator. By continually recording their estimates, many pupils could determine whether an answer was reasonable. We were then ready to continue writing our own story problems and creating a class problem-solving booklet.

Writing and Solving Problems

What was the final result of using calculators to solve pupil-generated problems? Once pupils had access to a calculator, their motivation increased and productivity was high. With the exception of one pupil, all chose to use the calculator at some point. One particularly bright pupil had written a problem to determine the number of students in the school. She added eight two-digit numbers on the calculator and was able to read the total, even though it went beyond the first-grade curriculum. A final draft, or "publishable form," of each pupil's own favorite problem was included in a class booklet. This booklet was continually signed out to be taken home and shared with parents. The pupils chose to include in our class booklet the following examples that were solved by using a calculator. The teacher and pupil worked together to produce the final published form from the pupil's original problem.

Original problem

> The Oilers scored 22 goals in 4 oams.
> They scored 92 goals all seasn.
> how many for The rest?

Final form

The Oilers scored 22 goals in 4 games. They scored 92 goals all season. How many goals did they score for the remainder of the season?

Original problem

> There are 48 Frog and 48 Dog
> How mane are toogox.

Final form

There are 48 frogs and 48 dogs. How many animals do we have altogether?

Original problem

> anna had 44 flower stikrs.
> She Gave 28 to me
> thay Were Pritly. thay Were
> Culerfule. Haw meny duse
> anna have lift?

Final form

Anna had 48 flower stickers. She gave 28 to me. They were pretty. They were colorful. How many does Anna have left?

Original problem

> One daY I went to the
> Store. I got 12 Jackets
> and five Baseball glovs and
> 10 comics.
>
>   the end

Final form

One day I went to the store. I got 12 jackets and 5 baseball gloves and 10 comics. How many things did I buy?

Summary

The use of the calculator allowed the focus of the lesson to be on the problem-solving process, not computation. With the aid of a calculator most pupils could comfortably work with whole numbers to 100 and were no longer restricted by difficulty of computation. They were motivated to write and solve more problems than would have been possible without a calculator. Since the pupils were not restricted to numbers less than 10 but could use large numbers, their problems became more "realistic" to them and more interesting applications resulted. In addition, pupils seemed to become more comfortable with estimation strategies before using the calculator. The ability to compute using a calculator definitely enhanced our primary-level class's problem-solving project.

Bibliography

Bruni, James V. "Problem Solving for the Primary Grades." *Arithmetic Teacher* 29 (February 1982):10-15.

Duea, Joan, George Immerzeel, Earl Ockenga, and John Tarr. "Problem Solving Using the Calculator." In *Problem Solving in School Mathematics*, 1980 Yearbook of the National Council of Teachers of Mathematics, edited by Stephen Krulik and Robert E. Reys, 117-26. Reston, Va.: The Council, 1980.

Fennell, Francis (Skip), and Richard Ammon. "Writing Techniques for Problem Solvers." *Arithmetic Teacher* 33 (September 1985):24-25.

Reys, Robert E., Barbara Bestgen, Terrence Coburn, Robert Marcucci, Harold Schoen, Richard Shumway, Charlotte Wheatley, Graysn Wheatley, and Arthur White. *Keystrokes: Calculator Activities for Young Students: Counting and Place Value.* Palo Alto, Calif.: Creative Publications, 1980.

The Calculator Is a Problem-solving Concept Developer
by Margaret Comstock and Franklin Demana

The hand-held calculator is a powerful problem-solving tool. It can be used to develop concepts and explore mathematical topics. The speed and power of the calculator make more realistic and interesting problems accessible and allow students to work many more problems than possible with pencil and paper. The calculator focuses students attention on mathematical processes. Numerical problem solving forms the foundation from which algebraic ideas grow. This article gives examples of calculator table-building activities suitable for middle school curriculum that develop problem-solving skills and mathematical concepts and span the gap from arithmetic to algebra.

The examples are taken from the materials developed in the "Approaching Algebra Numerically" (AAN) project at Ohio State University for seventh- and eighth-grade students (Demana, Leitzel and Osborne, in press). The AAN materials are dedicated to establishing the concepts of variable and function and are in their third year of use. The College Entrance Examination Board cited the calculator-based numerical problem-solving AAN materials as modules that exemplify those needed in a computational problem-solving course that they recommend be designed for students in ninth grade who are not ready to begin the traditional college preparatory sequence (James, 1985). The AAN materials grew out of an earlier project that successfully designed calculator-based materials for college-bound high school seniors with serious deficiencies in mathematics (Demana and Leitzel 1984; Leitzel and Osborne 1985).

Percent

Students have a good deal of trouble mastering this important topic in the middle school curriculum One reason is that students are not able to do a sufficient number of problems with pencil and paper to gain the experience necessary to master the concept. Calculators make it possible for students to do many more problems in the same period of time. We begin by having the student compute several special cases of a problem in tabular form. (See table 1.) Several important concepts can be developed with table 1 as a starting point.

1. Mathematical process

By noting the form in which entries are recorded in columns 2 and 3 of table 1, we can focus students' attention on the process used to arrive at an answer rather than on the answer itself. Teachers can emphasize the process by asking "What did you do to get your answer?" Students will see that the entry in the second column is 0.30 times the entry in the first column and that the last entry is the first minus the second The use of a number pattern like 20, 30, 40 in the first column helps students discover that the pattern for the last column is 0.70 times the entry in the first column or equivalently that the sale price is 70 percent of the original price.

Table 1 A 30-Percent-Off Sale at Sparks Department Store		
Original price ($)	Discount ($)	Sale price ($)
20	$0.30(20) = 6$	$20 - 6 = 14$
30		
40		

2. Rounding

If we add 39.96 to the first column in table 1 then the discount is $0.30(39.96) = 11.988$. Rounding to two decimal places is natural because the numbers in this problem represent amounts of money. However, this computation leads to healthy classroom discussion. Will the store round the discount to $11.99 or truncate it to $11.98? This entry, when considered from the store's point of view, demonstrates a reasonable situation where truncation may be preferred to rounding in the usual fashion. In addition, such numbers as 39.96 are likely not to be used when students work only with pencil and paper.

3. Problem Solving

Students can be introduced to numerical problem solving by adding additional lines to the completed

Table 2
Extending Table 1 to See More Patterns

Original price ($)	Discount ($)	Sale Price ($)
20	6	14
30	9	21
40	12	28
	13.50	24.50

version of table 1. (See table 2.) Students' attention should be focused on the number patterns in the table. For example, teachers can ask, "Using the entries in the table, can you predict the discount for an original price of $25?" You will be pleasantly surprised to find that most students will guess halfway between 6 and 9, or $7.50. The teacher can help students with the fourth line of the table by asking, "What original price will give a discount of $13.50?" Then the information in the table can be used to show that an original price greater than $40 is needed to obtain a discount of $13.50 on an item. Then by guessing and checking students quickly find that the original price must be $45. Some students may actually reverse the process and divide 13.50 by .3 to get 45 but we have found that most seventh-graders need to guess and check to solve this problem.

Table 3
Introducing the Concept of Variable

Original price ($)	Discount ($)	Sale Price ($)
20	0.30(20)	20 - 0.30(20)
30	0.30(30)	30 - 0.30(30)
40	0.30(40)	40 - 0.30(40)
x		

Finding an original price that gives a sale price of $24.50 can be done in a similar way. The entries in the third column of the table make it clear that the answer lies between $30 and $40. Again the student can guess and check to see that the original values for the discount and the sale price that are not halfway between the computed values should also be given. Tables can set the stage for problem solving activities by helping students sharpen their estimation skills.

4. Variables and equations

We can introduce the concepts of variable by adding another row to table 1. (See table 3.) The entries in the last two columns of table 3 are in a form that focuses students' attention on the process used to find the entries. Students are better able to see that the missing entries in the last line of table 3 are 0.30x and x-0.30x. In addition, this approach allows students to summarize (generalize) several special cases and write a model for a problem. Now students are ready to write equations.

Sparks Department Store is having a 30-percent-off sale. If x is the original price of the item, write an equation that says the sale price of an item is $50.

Solution: From the last entry in table 3 we see that $x - 0.30x = 50$ is the desired equation.

Our experience in using this approach with middle school students is that the concept of variable is natural for students when it is based on a generalization of situations that have previously been studied numerically.

5. Function

Table 1 gives values for two distinct functions; namely, 30 percent of a number and a number reduced by 30 percent. Such examples offer students extensive experience with the concept of of function in a numerical setting. As with the concept of variable, the notion of function is also more natural for students when it grows from situations that have already been studied numerically.

Area and Perimeter

Students frequently confuse area with perimeter. We have found that by having a table-building activity combining both ideas students are able to compare and contrast and thus overcome this confusion. Students' understanding is better when concepts are integrated in a given problem.

The following important concepts can be developed from table 4

1. Mathematical process

The computations in the third column of table 4 introduce ideas not encountered in table 1. Students

Table 4 Helping Students Distinguish Area and Perimeter

Width (m)	Length (m)	Perimeter (m)	Area (m²)
2	5	14	10
3.5			
5			

will find the perimeter of these rectangles in at least three distinct ways:

a) By adding the measures of all four sides
b) By adding the width and the length and then doubling the result
c) By doubling both the width and the length and then adding the results

Classroom discussion can focus on why these three processes lead to the same result. This discussion provides the student with numerical evidence that the algebraic expressions $W + L + W + L$, $2(W + L)$ and $2W + 2L$ are equal.

2. Accuracy

Completing the area column in the second row of table 4 leads to the computation $(3.5)(6.5) = 22.75$. Discussion can focus on whether 22.75 should be rounded to tenths. For example, if the width is considered accurate to tenths, then 22.75 should be rounded to 22.8. However, if the width is to be accurate to hundredths, then 22.75 is the best answer. Of course, the teacher can just give specific directions about rounding.

3. Problem solving

Once again we can focus on problem solving by adding additional lines to table 4. (See table 5.) Completing the fourth line of table 5 requires the students to reverse the process of adding 3 to the width to find the length. So if the length is 10, then the width is 7 and the perimeter and area are 34 and 70, respectively. Filling in the rest of the fourth line helps the students with the rest of the table. Students can now observe that the width needed for the fifth line is a little larger than 7. Then by guessing and checking the students can find that a width of 8 gives perimeter of 38. Similarly, the students can use the entries in the table to see that the width needed for the last line of the table is between 5 and 7. A quick computation shows the width to be 6.

4. Variables and equations

If we add another line to table 4 using W for the width, then the length will be $W + 3$ and the area $W(W + 3)$. Depending on the process the student uses to compute the perimeter, you will see such expressions as

$$W + W + 3 + W + W + 3$$
$$2(W + W + 3),$$
$$4W + 6$$

or

$$2W + 2(W + 3)$$

Another area problem is the following:

A rectangle has a length three meters more than its width. If W stands tor the width of the rectangle, write and solve an equation that states the area of the rectangle is 154 square meters.

Solution: By referring to the preceding discussion we see that $W(W + 3) = 154$ is the desired equation.

Table 5
Extending Table 4 to See More Patterns

Width (m)	Length (m)	Perimeter (m)	Area (m²)
2	5	14	10
3.5	6.5	20	22.75
5	8	26	40
	10		
		38	
			54

Although solving this equation is not part of the middle school curriculum, by guessing and checking and using calculator, middle school students will easily find the solution to be $W = 11$.

Exponents

Our middle school students enjoy problems about compound interest and population growth. These problems would certainly not be accessible to students without the aid of calculators. One such problem is this:

The population of Millersport now is 300,000. The population will increase at the rate of 4.5 percent each year for the next twenty years. Complete table 6.

Table 6
What Populations Can We Expect with a 4.5-Percent Increase Each Year?

Years elapsed	Population
0	300 000
5	
10	
15	
20	

Table 7 A Completed Table Can Be a Source of New Questions

Years elapsed	Population
0	300 000
5	373 855
10	465 891
15	580 585
20	723 514

This example also furnishes a natural situation in which rounding to whole numbers is appropriate and focuses on the following topics:

1. Mathematical process

To complete table 6 students must first see how to compute the population. The population after one year is

$$300,000 + 0.045(300,000)$$

or 300,000(1.045). This application of the distributive property is difficult but worthwhile to explain. Once they overcome this hurdle, students can see that the population two years later is

$$300,000(1.045)(1.045)$$

or $300,000(1.045)^2$ The population five years later will be $300,000(1.045)^5$

2. Problem solving

Table 6 supplies a basis for many interesting problems. (See table 7.)

When will the population of Millersport be 500,000? Students can observe in table 7 that the population will be 500 000 between ten and fifteen years from now. With a little more computation students will see that the population in eleven years is

$$300\ 000(1.045)^{11} = 486\ 856$$

and in twelve years.

$$300\ 000(1.045)^{12} = 508\ 764.$$

So the population will be 500 000 between eleven and twelve years from now. We found that our students were not satisfied with this answer and tried to find the exact number of years. The answer is

$$300\ 000(1.045)^{11.6} = 499885$$

Exponents other than whole numbers are quite natural when calculators are used. A very lively discussion can result if the teacher now asks. "In what month during the eleventh year will the population of Millersport reach 500,000?"

3. Variables and equations

If n stands for the number of years elapsed, then the population of Millersport n years later is $300\ 000(1.045)^n$. Middle school students could not solve equations with this expression without guessing and checking with a calculator. Students gain a real sense of power through working problems of this type.

Summary

We have used a calculator-based numerical problem-solving application to develop mathematical concepts with college freshmen, high school seniors and middle school students. Our experience has shown that calculators keep students' interest levels high and inspire them to work numerous challenging problems not possible with only pencil and paper. In addition, teachers employing this approach have been able to get students to verbalize about mathematics because the calculator stimulates students' talk about mathematical situations. Such a method also enhances students' understanding of mathematics and provides them with a firm grounding in mathematics necessary for living in a high-technology society.

Bibliography

Demana, Franklin D., and Joan R. Leitzel. *Transition to College Mathematics*. Reading, Mass.: Addison Wesley Publishing Co., 1984.

Demana, Franklin D., Joan R. Leitzel, and Alan Osborne. *Seventh and Eighth Grade Units—Approaching Algebra Numerically Project*. Lexington, Mass.: D. C. Heath & Co., in press.

Herbert, James, ed. *Academic Preparation in Mathematics*. New York:College Entrance Examination Board. 1985

Leitzel, Joan R., "Calculators Do More Than Compute." In *New Directions in Two-Year College Mathematics: Proceedings of the Sloan Foundation Conference on Two-Year College Mathematics*. New York: Springer-Verlag. 1985

Leitzel, Joan R., and Alan Osborne. "Mathematical Alternatives for College Preparatory Students." In *The Secondary School Mathematics Curriculum*, 1985 Yearbook, edited by Christian R. Hirsch and Marilyn J. Zweng. Reston, Va.: National Council of Teachers of Mathematics, 1985.

National Council of Teachers of Mathematics. *An Agenda for Action: Recommendations for School Mathematics of the 1980s*. Reston, Va.: The Council, 1980.

Teacher-made Overhead Manipulatives

by James H. Wiebe

We all have experienced the frustration of trying to use manipulatives to model some mathematics concept to an entire classroom of pupils. Pupil-sized manipulatives are too small for most of the class to see, and besides, how do you hold them up for all to see while manipulating them? Chalkboard drawings cannot be manipulated. Felt-board manipulatives are a little better, but they, too, may be difficult for students to see, can be difficult to manipulate, and require special materials.

To the rescue, your trusty overhead projector on which transparent models can be manipulated and projected for all to see! A limited number of transparent manipulative materials are available commercially, but they are expensive and limited to a few models (e.g., Cuisenaire rods). The alternative is to make transparent models yourself. These manipulatives are easy to make with readily available materials, are easy to manipulate, and are highly visible to everyone in the class. Furthermore, some evidence indicates that students can benefit just as much from watching the teacher manipulate models as from performing their own manipulations (Knaupp 1970).

The general steps for making overhead manipulatives are as follows:

1. Draw the desired shapes in black ink on white paper or use the shapes in the figures in this article.
2. Make a photocopy of the drawings (note: better transparencies result from photocopied drawings than from the original ink or pencil drawings).
3. Copy the photocopied drawings onto transparencies using a heat-transfer copy machine (e.g., Thermofax machine).
4. Cut out the transparent shapes.
5. Color the shapes, if needed, with permanent colored marking pens (e.g., Sharpie permanent markers).
6. You may wish to slightly curl up one or more corners of your manipulatives so that they can be easily picked up and moved on the overhead projector.

What kinds of overhead manipulatives should you make? Of course, that choice depends on the type of manipulatives you use in your class. Descriptions of some of the more useful manipulatives and suggestions for making them follow:

Base-ten blocks

Materials: The shapes in figure 1, heat-transfer transparencies, scissors.

Uses: Numeration, whole-number operations, decimal operations, metric measurement concepts.

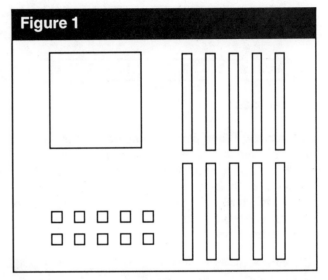

Figure 1

Procedure: Make two copies of the shapes in figure 1 on transparencies and cut them out.

Colored chips

Materials: A nickel or other object to assist in drawing circles, a black flow pen, paper, transparencies, permanent colored markers, scissors.

Uses: Numeration, whole-number operations, other bases (base five, etc.).

Procedure: Draw circles on paper, copy them onto transparencies, and color them using marking pens. An alternative is to copy the circles on colored transparencies.

Place-value charts

Materials: Permanent marker, a transparency, dried beans or similar objects to be used as markers.

Uses: Whole-number operations, numeration, decimal operations.

Procedure: Draw the place-value chart directly on the transparency or draw it on paper and copy it onto the transparency. Use the beans as markers on the place-value chart.

Figure 2

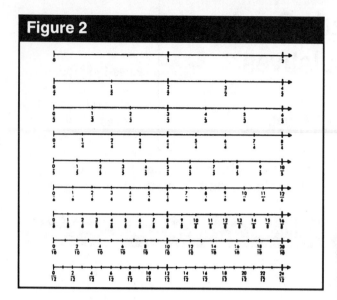

Number lines

Materials: Black pen and paper, a transparency, one-half-inch-wide strips of paper or one-half inch-wide strips of colored transparencies, scissors.

Uses: Numeration (e.g., rounding), whole-number operations, meaning of fractions, operations on fractions (see Wiebe [1988]), integers, decimal numeration, decimal operations, and so on.

Figure 3

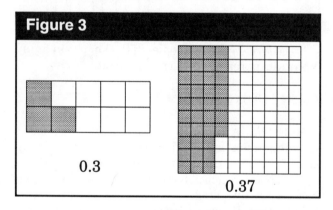

0.3

0.37

Procedure: On paper, draw number lines for wholes, halves, thirds, and so on, as shown in figure 2. The number lines should be parallel and to the same scale. Cut strips of appropriate length from colored transparencies or cut them from clear transparencies and color them with permanent markers. Use these strips to represent fractions, find equivalent fractions, and do operations on fractions. For a more detailed discussion of the use of number lines and strips to represent fractions and operations on fractions, see Wiebe (1985).

Grids

Materials: A black pen, paper, a transparency, colored permanent markers, scissors.

Uses: Fractions, decimals, and percentages

Procedure: Draw and shade shapes to represent fractions, decimals (e.g., a rectangle with 10 or 100 squares to represent tenths and hundredths, respectively [see fig. 3]), and percentages. Congruent squares can be shaded in different colors to represent equivalent fractions then superimposed to develop this concept (see fig. 4) (Wiebe 1988). Squares (10 cm by 10 cm) divided into 100 congruent squares and superimposed over other squares can be used to represent percentage problems (see fig. 5) (Wiebe 1986). For a detailed description of the use of transparencies in solving percentage problems, see Wiebe (1986).

Figure 4

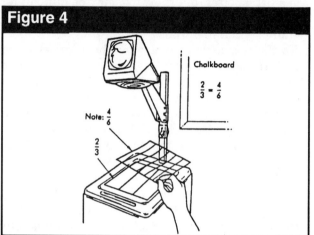

References

Knaupp, Jonathan Elmer. "A Study of Achievement and Attitude of Second-Grade Students Using Two Modes of Instruction and Two Manipulative Models for the Numeration System." Ph.D. diss., University of Illinois at Urbana- Champaign, 1970. *Dissertation Abstracts International* 31 (1971), no 6471A. (University Microfilms no. 71-14832).

Wiebe, James H. "Disccovering Fractions on a 'Fraction Table'." *Arithmetic Teacher* 33 (December 1985): 49-51.

——"Manipulating Percentages." *Mathematics Teacher* 79 (January 1986):23-26.

——*Teaching Elementary Mathematics in a Technological Age.* Scottsdale Ariz.: Gorsuch Scarisbrick Publishers, 1988.

Figure 5

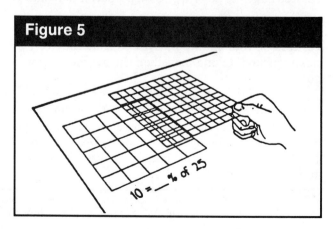

Index